Book of Tasty Stories

BTS

RECIPE BOOK 2

GLOBAL EDITION

Introduction

<BTS RECIPE BOOK 2> is a sequel to the much-loved <BTS RECIPE BOOK> that was released in 2022. Since the release of the <BTS RECIPE BOOK>, ARMY's cooking reviews have been shared across different social media sites with the hashtag #BTS_RECIPE_BOOK. Being able to cook and enjoy the same food that BTS made and ate has allowed ARMY to share more experiences with BTS. This sequel was created in response to the enthusiastic reaction from ARMY around the world.

This book maintains the key concept of the <BTS RECIPE BOOK>, as it **introduces the dishes created and eaten by BTS in various media, presents the BTS stories tied to each dish, and also provides BTS' own recipes.** Meanwhile, the number of dishes has increased compared to the previous book, and the overall cooking difficulty level of each dish has been set to low or middle. Therefore, with this book, even ARMY who is unfamiliar with cooking will easily be able to make a meal enjoyed by BTS.

Furthermore, the stories and recipes have been structurally divided for readers to enjoy each part to the fullest. The story section includes BTS stories associated with the dishes. **QR codes linked to the related videos** have been added to help provide the context of the stories for readers. Also, **a section offering explanations of simple Korean expressions related to each dish** (A Pinch of Korean) has been added to help readers better understand what BTS says or does. It will also be fun to learn about **BTS' personal tastes** (BTS Choice), which are included throughout the book, and compare them with your own tastes.

The recipe section is based on actual recipes used by BTS, while specific measurements, ingredients, and preparation steps were finalized after consulting with culinary experts. For ARMY who may have difficulty finding Korean ingredients, this book offers possible substitutes so that they can recreate the taste as close as possible. Also, when possible, **vegan** and **gluten-free** recipes are provided for each dish. All the dishes in this book are **nut-free**. However, since allergic reactions may vary from person to person, each reader should check for possible allergens in each of the ingredients and the possibility of cross-contamination before preparing any of the dishes. Lastly, this book provides **BTS' own cooking tips** (Special Tip) that readers can use to make the dishes even more delicious.

No matter how the completed dishes look or taste, BTS enjoys them with happy smiles on their faces. We hope that those who read this book can also relish the cooking process and enjoy the dishes with great pleasure. Treat your loved ones, or treat yourself, to a delicious meal cooked with love. **And once again, we hope you enjoy another delicious day with BTS!**

Cake

<BTS RECIPE BOOK 2>는 2022년에 출시되어 많은 사랑을 받은 <BTS RECIPE BOOK>의 후속편이다. <BTS RECIPE BOOK> 출시 후, 각종 SNS에서는 #BTS_RECIPE_BOOK 해시태그와 함께 아미들의 생생한 요리 후기가 공유되어 왔다. 방탄소년단이 만들고 먹은 음식을 아미들이 직접 만들어 즐김으로써 방탄소년단과 더욱 많은 경험을 공유할 수 있게 된 것이다. 이 후속편은 세계 각지의 아미들이 보내 준 뜨거운 반응에 응답하기 위해 제작되었다.

이 책은 방탄소년단이 여러 콘텐츠에서 직접 만들고 먹은 요리를 소개하며, 각 요리에 얽힌 방탄소년단의 이야기와 그들만의 레시피를 자세하게 풀어낸다는 점에서 <BTS RECIPE BOOK>의 주요 콘셉트를 그대로 유지했다. 한편, 요리의 가짓수는 이전 책에 비해 늘어났으며, 각 요리의 조리 난이도는 낮거나 중간 정도로 더 쉽게 구성하였다. 따라서 요리가 익숙하지 않은 아미도 이 책과 함께라면 방탄소년단이 즐긴 한 끼 식사를 거뜬히 만들어 낼 수 있을 것이다.

나아가, 이야기 영역과 레시피 영역을 구조적으로 분리하여 각각을 충분히 즐길 수 있도록 했다. 이야기 영역에서는 요리와 연관된 방탄소년단의 이야기를 다룬다. 이야기 속 상황을 쉽게 이해할 수 있도록 **각 영상으로 연결되는 QR코드**를 추가했으며, **각 요리 관련 간단한 한국어 설명을 제공하는 코너**(A Pinch of Korean)를 만들어 영상 속 방탄소년단의 말과 행동을 보다 자세히 이해할 수 있도록 도왔다. 곳곳에 수록된 **방탄소년단의 음식 취향**(BTS Choice)과 여러분의 취향을 비교해 보아도 재미있을 것이다.

레시피 영역은 기본적으로 방탄소년단의 레시피를 바탕으로 정리되었으며, 재료의 정확한 분량 등 상세한 조리법은 요리 전문가의 의견을 얻어 작성되었다. 한국 식재료를 구하기 어려운 아미를 위해 대체 가능한 재료를 표기하여 최대한 비슷한 맛을 낼 수 있도록 했고, 각 요리를 **비건, 글루텐 프리**로 조리할 수 있는 경우 그 조리법을 함께 안내했다. 또한, 이 책의 모든 요리는 **견과류 프리**다. 단, 알레르기 반응 여부는 개인에 따라 차이가 있으므로 조리 전 반드시 각 재료의 원료 및 교차 오염 가능성을 확인할 것을 권한다. 마지막으로 **방탄소년단의 요리 팁**(Special Tip)을 제공하여, 그들의 안내를 따라 요리를 더 맛있게 완성할 수 있도록 했다.

요리가 어떤 모양과 맛으로 완성되었든, 방탄소년단은 웃는 얼굴로 기꺼이 함께 즐긴다. 이 책을 보는 여러분도 요리를 만들고 먹는 과정에 즐거움이 가득하기를 바란다. 정성껏 만든 요리로 사랑하는 사람, 혹은 사랑스러운 자기 자신을 대접한다면 더할 나위 없겠다. **그렇게 다시 한번 방탄소년단과 함께하는 맛있는 하루가 되기를!**

Cake

Contents

		STORY	RECIPE
	Introduction	4	
	BTS' Kitchen		
	Kitchenware	8	
	Basic Ingredients	10	
	Spices, Seasonings, and Oils	12	
	Measurements	14	
1	**Easy Homemade Meals 간편한 집밥**		
	Ganjang Gyeran Bap (Egg Rice with Soy Sauce) 간장계란밥	20	164
	Kimchi Jjigae (Kimchi Stew) 김치찌개	28	166
2	**Box Lunches Just for You 취향 따라 만드는 도시락**		
	Gimbap (Seaweed Rice Rolls) 김밥	38	170
	Toast (j-hope's & Jung Kook's Egg Sandwiches) 토스트	46	174
3	**Slurp! Noodle Dishes 후루룩 면 요리**		
	Kalguksu (Noodle Soup) 칼국수	56	178
	Jjajang Ramyeon (Instant Noodles in Black Bean Sauce) 짜장 라면	64	182
4	**Spicy Meat Dishes 매콤한 고기 요리**		
	Dakgalbi (Spicy Stir-Fried Chicken) 닭갈비	74	186
	Jeyuk Bokkeum (Spicy Stir-Fried Pork) 제육볶음	82	190

		STORY	RECIPE
5	**Dishes for Special Feasts 특별한 만찬**		
	Miyeokguk (Seaweed Soup) 미역국	92	194
	Japchae (Stir-Fried Glass Noodles and Vegetables) 잡채	100	196
	Tteokguk (Sliced Rice Cake Soup) 떡국	108	198
	Bulgogi 불고기	116	200
6	**Perfect Pairings with Drinks 술과 어울리는 안주**		
	Kimchijeon (Kimchi Pancake) 김치전	126	206
	Gyeran Mari (Rolled Omelet) 계란말이	134	208
7	**Sweet Desserts 달달한 디저트**		
	Dalgona (Sugar Candy) 달고나	144	212
	Hwachae (Punch) 화채	152	214
	Basic Preparations 기본 조리법		
	Rice 밥		218
	Vegetable Broth 채소 국물		219
	Anchovy Broth 멸칫국물		220
	Beef Broth 고깃국물		221
	Ingredients Index	222	
	References	230	

Kitchenware

These are the tools often used in the recipes of this book.
이 책에 나오는 요리에 자주 사용되는 도구들을 소개한다.

Measuring spoons
계량스푼

Pot
냄비

Measuring cup
계량컵

Strainer
체

Skillet
프라이팬

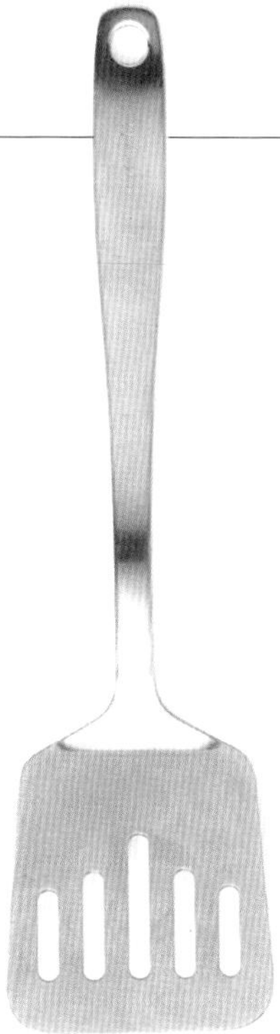

Spatula
뒤집개

Ladle
국자

Knife
칼

Bowl
볼

Paper towels
키친타월

Wooden chopsticks
조리용 나무 젓가락

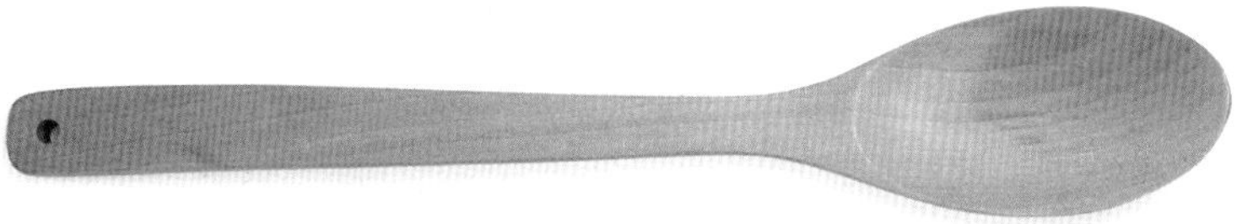

Wooden spoon
조리용 나무 주걱

BTS' *Kitchen*

Basic Ingredients

These are some of the basic ingredients commonly used in the recipes of this book. Most of them are easily found in Korean grocery stores, but if any of them are unavailable, check out the tips on the ingredients page of each recipe to see if they can be left out or replaced by something else.

이 책에 나오는 요리에 자주 사용되는 재료들을 소개한다. 대부분 한국 식료품점에서 쉽게 구할 수 있지만, 구하기 어려운 재료가 있다면 각 요리의 재료 페이지 팁을 참고하여 재료를 생략하거나 대체해도 된다.

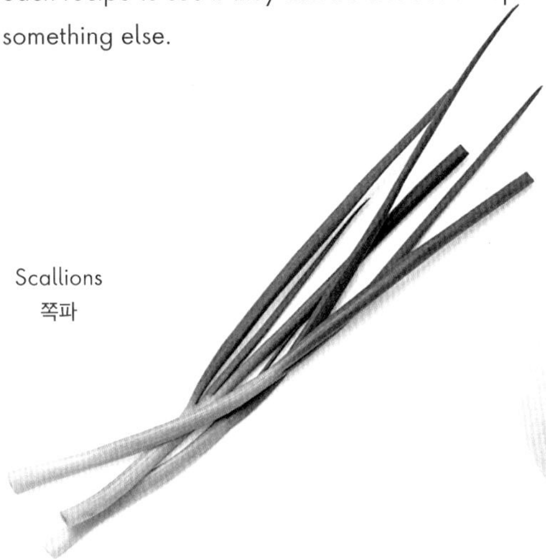

Scallions
쪽파

Onions
양파

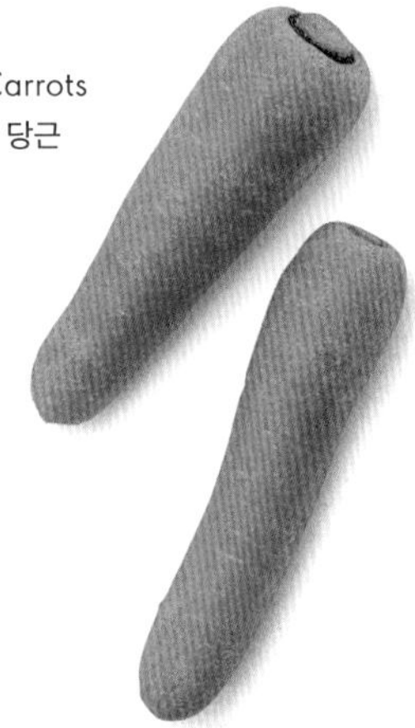

Carrots
당근

Large green onions (*Daepa*)
대파

*Though they look like leeks, they taste and smell more like green onions or scallions. If large green onions are unavailable, you can replace them with scallions, but note that scallions may have a weaker flavor.

Kimchi
김치

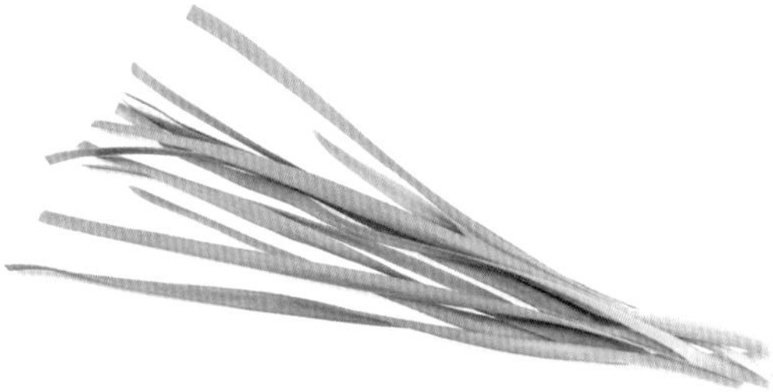

Chives
부추

Hot chili peppers
청양고추

Mushrooms
버섯

Eggs
계란

All-purpose flour
밀가루(중력분)

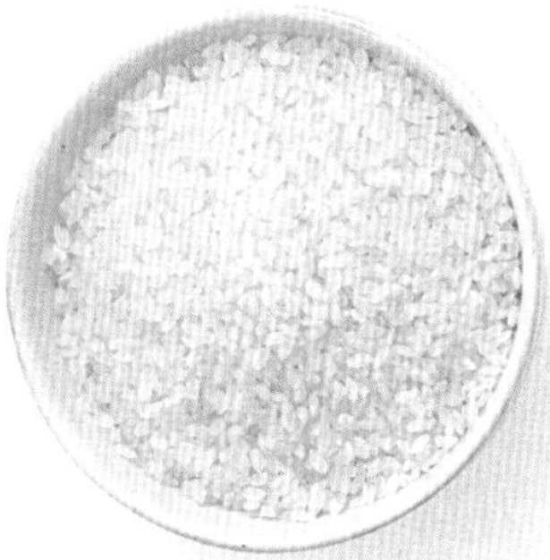

Uncooked rice
쌀

Dried anchovies for stock
국물용 멸치

* 2 to 3-in / 5 to 10-cm long dried anchovies
with heads and intestines removed

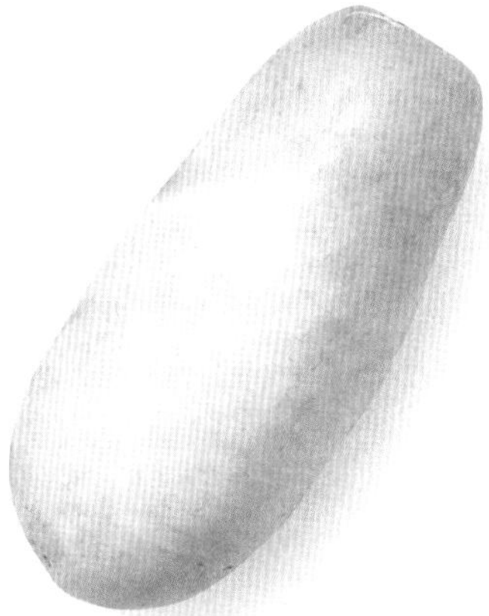

Korean radish
무

Tofu (Bean curd)
두부

Dried kelp
건조 다시마

Beef
소고기

Spices, Seasonings, and Oils

Gochutgaru
(Korean chili powder)
고춧가루

Toasted sesame seeds
참깨

Gochujang
(Korean red chili paste)
고추장

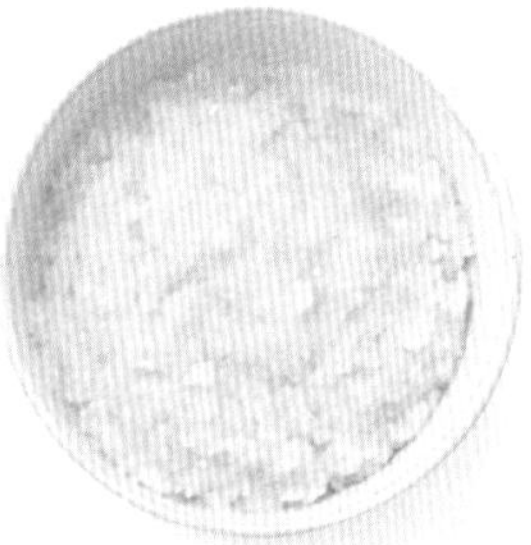

Minced garlic
다진 마늘

Soy sauce
간장

Soy sauce for soup
국간장

Sugar
설탕

Salt
소금

Cheongju
(Refined rice wine)
청주

Cooking oil
식용유

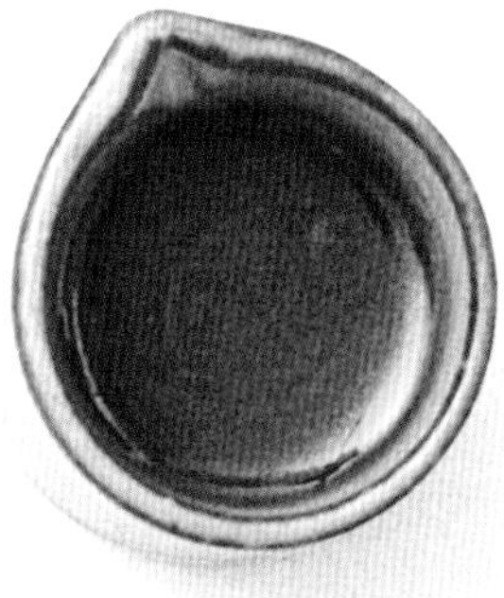

Toasted sesame oil
참기름

Black pepper
후추

Measurements

1 teaspoon 작은술

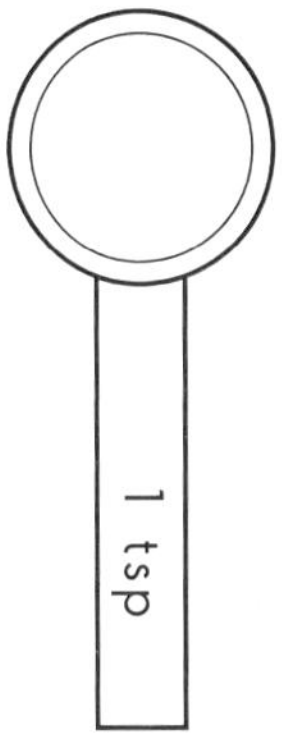

0.2 oz ≒ 5 ml

1 tablespoon 큰술

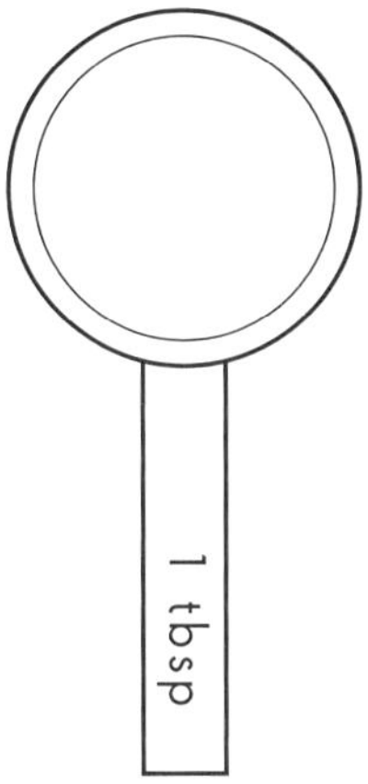

0.5 oz ≒ 15 ml

When measuring with a spoon, level off the top for dry and pasty ingredients, and fill to the brim for liquid ingredients.

스푼 계량 시 가루류와 장류는 윗부분을 평평하게 깎아 담고,
액체류는 가득 채워 담는다.

1 cup 컵

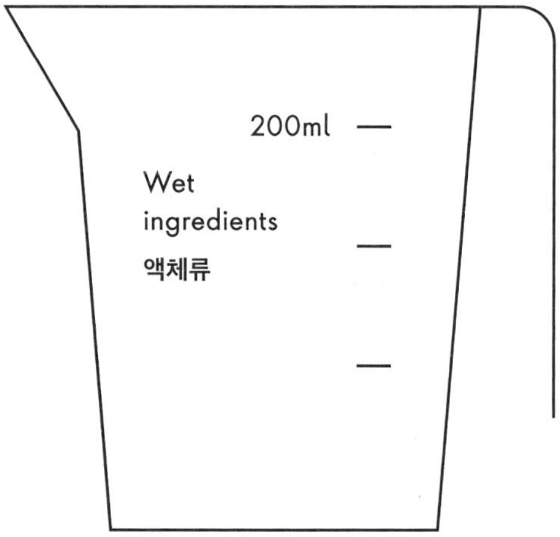

1 cup water
물 1컵
7 oz ≒ 200 ml

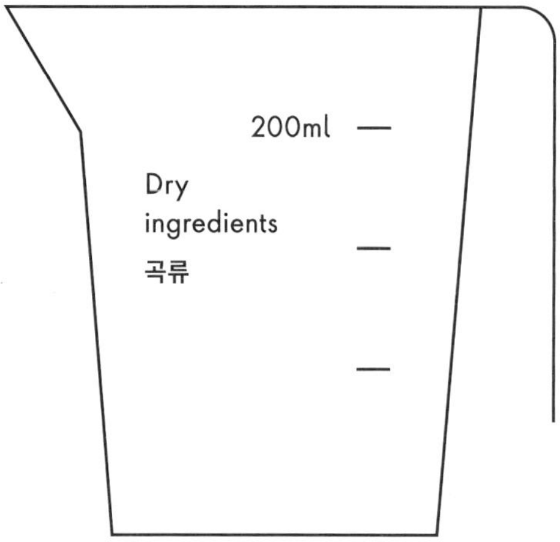

1 cup uncooked rice
쌀 1컵
6.3 oz ≒ 180 g

1 cup wheat flour
밀가루 1컵
4.5 oz ≒ 130 g

The standard volume of a cup is 8 oz / 250 ml in the US and 7 oz / 200 ml in Korea and Japan. In this book, one cup is 7 oz / 200 ml.

한 컵의 US 표준은 일반적으로 250ml, 한국과 일본 표준은 200ml이다. 이 책의 한 컵은 200ml를 기준으로 작성되었다.

* One cup was 8 oz / 250 ml in the <BTS RECIPE BOOK> released in July, 2022, so there is a difference in measurement between the two books.

2022년 7월에 출시된 <BTS RECIPE BOOK>에서는 한 컵이 250ml를 기준으로 작성되었으므로, 두 제품 간 컵 계량에 차이가 있음.

1

Easy Homemade Meals

간편한 집밥

Ganjang Gyeran Bap 간장계란밥

Kimchi Jjigae 김치찌개

- EASY HOMEMADE MEALS -

Ganjang Gyeran Bap

Egg Rice with Soy Sauce

간장계란밥

#Savory #Fried #Rice

RECIPE ⟶

page 164

"Bon voyage! Bon voyage! Bon voyage!" BTS chanted the theme song as they sat down to eat their last dinner in New Zealand on <BTS BON VOYAGE Season 4>. The dinner menu featured *ganjang gyeran bap* (egg rice with soy sauce) and *golbaengi somyeon* (spicy sea snails with thin noodles) made by j-hope, the leader of the younger members, and 95z (Jimin and V), as well as *tteokbokki* (stir-fried rice cakes) made by RM. *Ganjang gyeran bap* can be made easily with only cooked rice, eggs, soy sauce, and toasted sesame oil. SUGA recalled his mother making *ganjang gyeran bap* for him, and j-hope recalled his grandmother doing the same. Then they reminisced about their childhoods.

"본보야지! 본보야지! 본보야지!" 로고송을 부르며 시작된 <BTS 본보야지 시즌 4> 뉴질랜드 편의 마지막 저녁 식사. 이날 저녁은 막내 라인의 대장인 제이홉과 구오즈(지민과 뷔)가 만든 간장계란밥과 골뱅이 소면, 그리고 RM이 만든 떡볶이였다. 이 중 간장계란밥은 밥과 계란, 간장, 참기름만 있으면 매우 쉽게 만들 수 있는 요리다. 슈가는 어머니가, 제이홉은 할머니가 만들어 주셨던 간장계란밥을 떠올리며 어린 시절을 회상하기도 했다.

<BTS BON VOYAGE Season 4> EP. 8

Jimin and j-hope cook together 요리하는 지민과 제이홉

The last dinner in New Zealand 뉴질랜드에서의 마지막 저녁 식사

V

Can I fry the rest of the eggs?
I just want them over easy.

나머지 계란 내가 부쳐도 돼? 너 반숙으로 안 했지?

Jimin

They're sort of over easy.
Want to see?

나름 반숙인데. 볼래?

<BTS BON VOYAGE Season 4> EP. 8

Ganjang gyeran bap's recipe is so simple that you can cook it anytime without difficulty. Just put fried eggs on cooked rice, add soy sauce and toasted sesame oil, and mix them all together. If you fry the egg like the over-easy master V, the runny yolk will coat the grains of rice and enrich the flavor. To top it all off, by using the Korean-style soy sauce and toasted sesame oil recommended by RM and SUGA, you will be able to recreate the same nostalgic taste that BTS used to enjoy.

간장계란밥은 조리과정 역시 단순해 아무 때나 부담 없이 만들어 먹을 수 있다. 쌀밥에 계란프라이를 올려 간장과 참기름을 넣고 비벼 주면 완성이다. 이때, 반숙 마스터 뷔처럼 계란프라이의 노른자가 살짝 흐를 정도로 구워주면 노른자가 밥알을 코팅하여 풍미가 깊어진다. 여기에 RM과 슈가의 팁처럼 한국식 간장과 참기름을 사용한다면 방탄소년단이 먹었던 추억의 맛을 그대로 구현할 수 있을 것이다.

SUGA shares the *ganjang gyeran bap* recipe 간장계란밥 레시피를 알려 주는 슈가

The last night in New Zealand is filled with endless laughter 웃음이 끊이지 않는, 뉴질랜드에서의 마지막 밤

When an ARMY on Weverse once asked V what Korean food he would like to recommend to them, he replied, "do u know *ganjang gyeran bap*? *chamgireum* many many*." Adding *chamgireum* (toasted sesame oil) will make the dish twice as savory, so feel free to add more toasted sesame oil to your taste.

위버스에서 한 아미가 자신에게 추천해 주고 싶은 한국 요리가 있다면 무엇인지 뷔에게 묻자, 뷔는 "do u know ganjang gyeran bap? chamgireum many many (간장계란밥을 아시나요? 참기름 많이 많이)"라는 답글을 달았다. 뷔의 말처럼 참기름을 많이 넣으면 고소한 맛이 배가 되니 기호에 맞게 참기름을 추가해 보자.

---

* many many: The word "a lot" in Korean is ma-ni (많이). If you want to emphasize "a lot," you can say it twice consecutively. Ma-ni ma-ni (많이 많이) sounds similar to "many many" in English.

V

do u know ganjang gyeran bap?
간장계란밥을 아시나요?

chamgireum many many
참기름 많이 많이

Weverse (2020. 02. 05)

RECIPE →
page 164

<BTS BON VOYAGE Season 4> EP. 8

V and Jimin show off their cooking skills 요리 실력을 발휘하는 뷔와 지민

BTS gathers around the table 식탁에 옹기종기 모여 앉은 방탄소년단

j-hope checks if the rice is cooked well 밥이 잘 지어졌는지 확인하는 제이홉

<BTS BON VOYAGE Season 4> EP. 8

BTS has dinner together　함께 식사하는 방탄소년단

BTS *Choice*

Q.

Which dish would you recommend to ARMY who wants to make Korean food at home?

집에서 한식을 만들어 먹고 싶은 아미에게 추천해 주고 싶은 요리는?

Weverse (2020. 02. 05)

V

Ganjang gyeran bap 간장계란밥

A Pinch *of* Korean

완숙 *and* 반숙

Over-Hard and Over-Easy Eggs

While there are many expressions indicating the level of doneness of eggs in English, in Korean it is relatively simple. An egg is either wan-suk (완숙; fully cooked) or ban-suk (반숙; half cooked). Wan (완) of wan-suk means "fully," and suk (숙) means "cooked," so wan-suk refers to eggs where the yolk is fully cooked. Meanwhile, ban (반) of ban-suk means "half," so it refers to eggs with a yolk that is half-cooked, or runny. If you want to become a ban-suk master like V, check out the tips on page 165.

- EASY HOMEMADE MEALS -

Kimchi Jjigae

Kimchi Stew

김치찌개

#Spicy #Boiled #Stew

RECIPE →

page 166

Jimin and RM (MiniMoni), neither of whom are familiar with cooking, were full of worries before the cooking competition. Their mission that day for <Run BTS!> was to become SUGA's avatars and make *kimchi jjigae* (kimchi stew) and *kimchijeon* (kimchi pancakes). In Korea, where steamed rice is the staple food, people enjoy eating rice with stew, and *kimchi jjigae* is one of the most popular stews in the country. Triumphantly shouting "Let's do it!" MiniMoni started cooking. There was an instance where they confused salt and sugar, but fortunately they realized their mistake in time and were able to fix the situation.

요리에 서툰 지민과 RM(미니모니)은 요리 대결을 앞두고 걱정이 가득했다. 이날 <달려라 방탄> 미션은 슈가의 아바타가 되어 김치찌개와 김치전을 만드는 것이었다. 밥이 주식인 한국에서는 밥과 함께 찌개를 자주 먹는데, 김치찌개는 가장 흔하게 먹는 찌개 중 하나다. "께릿!"을 외치며 의기양양하게 시작했던 미니모니. 중간에 설탕과 소금을 구분하지 못해 실수할 뻔했으나, 다행히 늦지 않게 알아채고 수습할 수 있었다.

<Run BTS!> EP. 103

RM and Jimin prepare to cook 요리 시작 전 준비하는 RM과 지민

The rest of the members watch RM and Jimin cook RM과 지민이 요리하는 장면을 지켜보는 나머지 멤버들

Jimin

I'm so nervous...
I'm really not a good cook, you know...

여기 막상 오니까 무섭다... 내가 요리를 잘 못하니까...

RM

Jimin, imagine how scared I must be.

지민아, 난 얼마나 무섭겠니?

<Run BTS!> EP. 103

Making *kimchi jjigae* is simpler than you think. Just like in SUGA's recipe, you just boil kimchi, meat, and a variety of vegetables together and season it with salt and sugar. Depending on the ingredients you add, there are many types of *kimchi jjigae*, including pork *kimchi jjigae*, tuna *kimchi jjigae*, and saury *kimchi jjigae*. BTS' *kimchi jjigae*, however, always has a generous amount of meat.

김치찌개 만드는 법은 생각보다 간단하다. 슈가의 레시피처럼 김치와 고기, 각종 채소를 넣어 끓인 후, 소금과 설탕 간을 해 주면 완성이다. 넣는 재료에 따라 돼지고기 김치찌개, 참치 김치찌개, 꽁치 김치찌개 등 김치찌개 종류가 다양한데, 방탄소년단의 김치찌개에는 언제나 고기가 가득하다.

Kimchi jjigae is served in a squirrel bowl that BTS loves
방탄소년단이 좋아하는 다람쥐 그릇에 담긴 김치찌개

Jung Kook smiles as he watches RM and Jimin cook   RM과 지민이 요리하는 모습을 보며 웃는 정국

RM accidentally adds minced garlic in the *kimchijeon* batter instead of the *kimchi jjigae*
김치찌개에 넣어야 할 다진 마늘을 실수로 김치전 반죽에 넣은 RM

SUGA

I think I added more meat
(to this kimchi jjigae)
than meat soup.

(김치찌개에) 고깃국보다 고기를 더 많이 넣은 것 같아.

<BTS BON VOYAGE Season 4> EP. 4

Even on their trip to New Zealand, SUGA made *kimchi jjigae* so abundant in pork that it made BTS' eyes widen. The hot, hearty stew was perfect for melting away the cold and relieving the fatigue of BTS, who had worked hard all day. Anyone who tastes the fantastic combo of SUGA's *kimchi jjigae* and steaming rice will exclaim, "This is so delicious!" If the rice and stew isn't enough, how about adding sausage and *ramyeon* noodles (instant noodles) to the leftover broth for a second round like BTS did?

모두 함께 떠난 뉴질랜드 여행에서도 슈가는 멤버들의 눈이 휘둥그레질 정도로 돼지고기를 듬뿍 넣어 김치찌개를 완성했다. 기름지고 따뜻한 국물은 하루 종일 바삐 움직인 방탄소년단의 추위와 피로를 풀어 주기에 충분했다. 슈가표 김치찌개와 따뜻한 밥의 환상적인 조화를 맛본다면 누구라도 "아이 잘 먹었다!" 소리가 절로 나올 것이다. 밥과 찌개로 부족하다면, 방탄소년단처럼 남은 국물에 소시지와 라면 사리를 추가해 2차전을 즐겨 보는 것은 어떨까?

RECIPE →
page 166

BTS eats *kimchi jjigae* together　사이좋게 김치찌개를 나눠 먹는 방탄소년단

<Run BTS!> EP. 103

Jin, V, and Jung Kook taste the food 요리를 맛보는 진, 뷔와 정국

BTS *Choice*

Q.

Which dish would you recommend to ARMY who wants to make Korean food at home?

집에서 한식을 만들어 먹고 싶은 아미에게 추천해 주고 싶은 요리는?

<BTS RECIPE BOOK>

Jin

Kimchi jjigae 김치찌개

Jung Kook

Pork *kimchi jjigae* 돼지 김치찌개

A Pinch of Korean

Changing Declarative Sentences into Imperative Sentences

In episode 103 of <Run BTS!>, Jin and SUGA were in charge of directing and giving orders to the other members, whose job was to cook. One way to change declarative sentences into imperative sentences in Korean is to switch the ending "-다" with "-아 / -어." However, imperative expressions like these are not honorifics, so it is only acceptable to use them among close friends. It can be considered rude when you say them to someone you meet for the first time or someone older than you, so be careful when using these expressions.

Ingredients/Dish		Basic Verb Forms		Imperative Sentences
ma-neul 마늘 garlic	+	*ppa-ta* 빻다 to grind	=	*ma-neul ppa-a* 마늘 빻아. Grind the garlic.
go-gi 고기 meat	+	*neo-ta* 넣다 to put in	=	*go-gi neo-eo* 고기 넣어. Put the meat in.
jji-gae 찌개 stew	+	*kkeu-ri-da* 끓이다 to boil	=	*jji-gae kkeu-ryeo* 찌개 끓여. Boil the stew.

2

Box Lunches Just for You

취향 따라 만드는 도시락

Gimbap 김밥

Toast 토스트

- BOX LUNCHES JUST FOR YOU -

Gimbap

Seaweed Rice Rolls

김밥

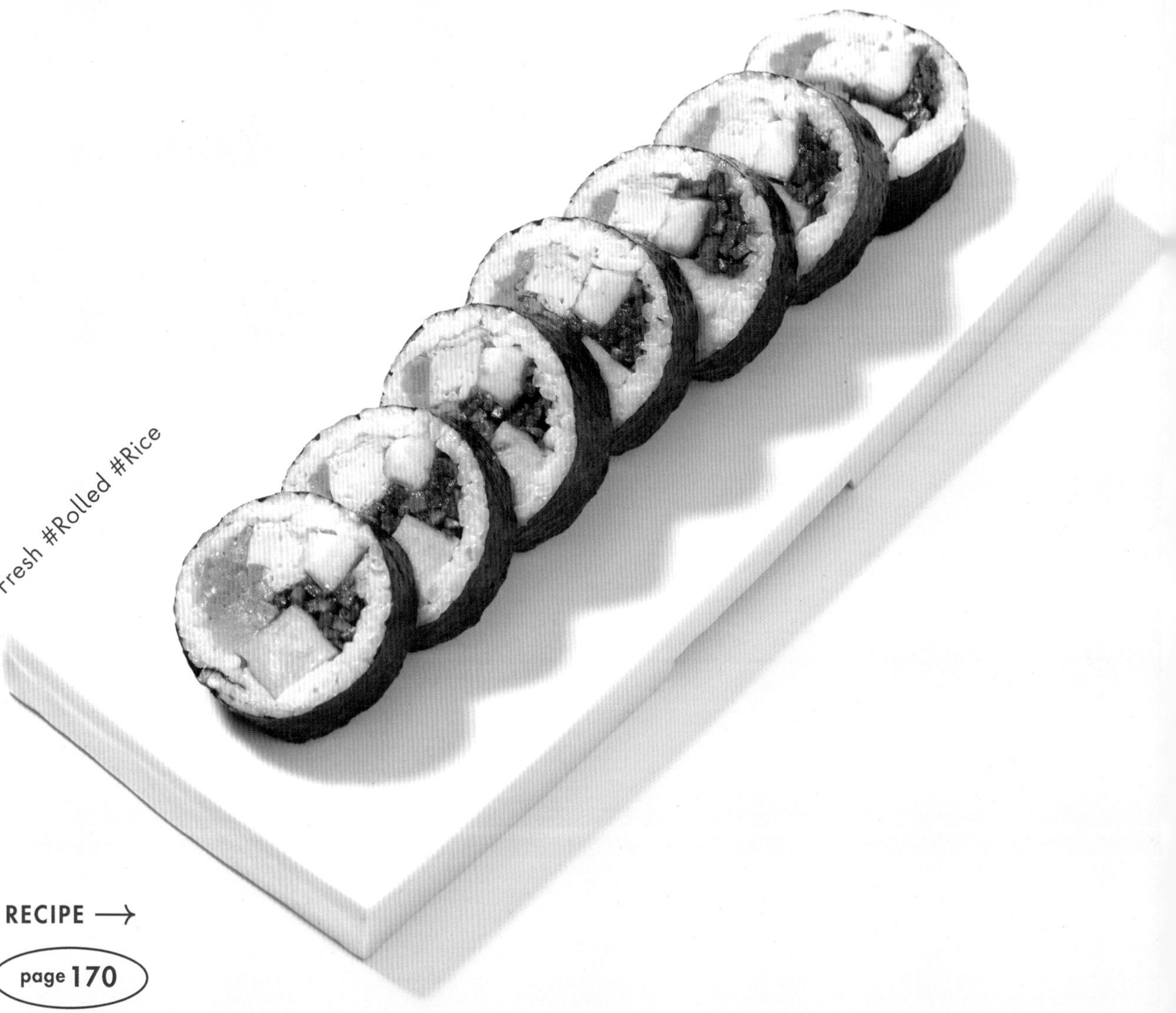

RECIPE →

page 170

A determined-looking Jimin and Jung Kook appeared on a live broadcast one summer day. At the beginning of the broadcast, the two sat without talking. Jimin stared intensely into the camera, and Jung Kook struggled to hold back his laughter. In front of them were ingredients for none other than *gimbap* (seaweed rice rolls)! On this live broadcast, Jimin and Jung Kook made *gimbap* as they interacted with ARMY. In Korea, *gimbap* is a common lunchbox food for picnics and outings. Jimin said that when he was young, *gimbap* was a must-have for school picnics and hikes with his father. This reminded Jung Kook of the 30 rolls of kimchi *gimbap* that his mother used to make for him when he was younger.

어느 여름의 라이브 방송에 비장한 모습의 지민과 정국이 등장했다. 방송 시작과 동시에 둘은 한동안 아무 말 없이 카메라를 바라봤다. 강렬한 눈빛을 쏘는 지민과 그 옆에서 웃음을 참는 정국의 앞에 놓인 것은 다름 아닌 김밥 재료! 이날 지민과 정국은 아미와 소통하면서 김밥을 만들었다. 한국에서 김밥은 소풍이나 나들이에 자주 등장하는 도시락 메뉴다. 지민은 어렸을 때 학교에서 소풍을 가거나 아버지와 등산을 하면 김밥을 필수로 챙겨갔다고 했다. 이에 정국도 어릴 때 어머니가 30줄씩 싸 주시던 김치김밥을 떠올렸다.

Jimin

It's been a while
since I made gimbap.
I have a feeling
I will make it really well today.

진짜 오랜만에 김밥을 만들거든요.
오늘 되게 잘 만들 수 있을 거라는 생각이 드네요.

Weverse (2020. 06. 21)

EP. 1

EP. 2

Weverse (2020. 06. 21)

Jimin waits for Jung Kook to help him put on plastic gloves 정국이 비닐장갑을 끼워주길 기다리고 있는 지민

Jung Kook and Jimin work hard making *gimbap* 열심히 김밥을 만드는 정국과 지민

Once the *gimbap* ingredients are ready, you're already halfway done. Now all you have to do is spread a thin layer of cooked rice seasoned with toasted sesame oil and salt on top of a dried seaweed sheet. Then arrange the ingredients on the rice, roll it up, and cut it into bite-sized pieces. There are many variations of *gimbap* with different ingredients, such as tuna *gimbap*, pork belly *gimbap*, cheese *gimbap*, and vegetable *gimbap*. Fill a lunchbox with your very own *gimbap* containing your favorite fillings, grab some chopsticks, and you're ready for a picnic!

On <IN THE SOOP BTS ver. Season 1>, RM and j-hope also made *gimbap* to take on their hike. Although the sides of the *gimbap* burst open while they were making it, it still tasted incredible. If the weather is nice, why not make some *gimbap* and head outdoors like RM and j-hope? The beautiful scenery will be the perfect side dish for your *gimbap* and make it taste even better.

김밥은 재료가 준비되었다면 이미 반은 완성이다. 우선 참기름과 소금으로 간을 한 밥을 김 위에 얇게 펴고, 밥 위에 재료를 넣고 돌돌 말아서 먹기 좋은 크기로 잘라 주면 끝이기 때문이다. 김밥은 재료에 따라 참치김밥, 삼겹살 김밥, 치즈김밥, 야채김밥 등 종류도 다양하다. 원하는 재료로 가득 채운 나만의 김밥을 도시락 통에 담고 젓가락만 챙기면 소풍 준비 완료!

<인더숲 BTS편 시즌 1>에서 RM과 제이홉도 등산할 때 가져갈 도시락으로 김밥을 만들었는데, 비록 만드는 과정에서 김밥이 터져버렸지만 맛은 훌륭했다. 날씨가 좋다면 RM과 제이홉처럼 김밥을 싸서 야외로 떠나 보자. 풍경을 반찬 삼아 먹는 김밥의 맛은 더욱 근사할 것이다.

RECIPE →
page 170

<IN THE SOOP BTS ver. Season 1> EP. 6

RM and j-hope fill a lunchbox with *gimbap* for their hike 산에 가져갈 김밥 도시락을 싸는 RM과 제이홉

RM

The scenery whets my appetite for gimbap.

풍경 한번 보니까 김밥이 쑥쑥 들어간다.

j-hope

They may not look perfect, but they're delicious!

모양은 이래도 맛은 있어!

BTS *Choice*

Q.

How much rice should be used in order to make delicious *gimbap*?

김밥에는 밥이 얼만큼 들어가야 맛있을까?

Weverse (2020. 06. 21)

Jung Kook

I don't like *gimbap* made with a lot of rice.
저는 밥이 많은 걸 별로 안 좋아해요.

Jimin

No, you need more rice than you think. To everyone watching this, *gimbap* needs to have a good amount of rice. That way, you can enjoy the fullness of the rice in your mouth.
아니야, 밥이 생각보다 많이 들어가야 해.
여러분, 김밥은 밥 양이 어느 정도 되어야지 입안 가득 차는 맛을 느낄 수 있어요.

A Pinch *of* Korean

옆구리 터진 김밥

Korean Expressions Related to *Gimbap*

When Jimin cut the *gimbap*, its side burst. When it is torn like this, the *gimbap* is called "yeop-gu-ri (옆구리; side) teo-jin (터진; burst) *gimbap*." And the two ends of a roll of *gimbap* are called kkong-da-ri (꽁다리; edge). These pieces are not usually included in box lunches, as they are shaped unevenly. However, they are actually the most delicious, as they have more of the ingredients in them than the other parts. So make sure to try them as you make your *gimbap*!

- BOX LUNCHES JUST FOR YOU -

Toast

j-hope's & Jung Kook's Egg Sandwiches

토스트

RECIPE →

pages 174-175

One peaceful morning on <IN THE SOOP BTS ver. Season 2>, Jung Kook woke up first and whipped up some golden toast (an egg sandwich) with eggs, cheese, and ketchup. Jin got out of bed just in time and had the first chance to taste Jung Kook's toast. "Oh, it's hot!" he cried out as he took a bite, but that did not stop him from going back for more. Jung Kook's toast was too delicious to stop at a single bite!

<인더숲 BTS편 시즌 2> 속 어느 평화로운 아침, 가장 먼저 하루를 시작한 정국은 순식간에 계란, 치즈와 케첩을 넣은 황금빛 토스트를 만들었다. 때마침 일어난 진은 정국이 만든 토스트를 첫 번째로 맛볼 기회를 얻었다. 진이 첫입에 "앗, 뜨거워!"를 외치면서도 또 한입을 베어 문 것을 보면 정국표 토스트가 한 입만 먹기에는 아쉬울 만큼 맛있었던 것이 분명하다.

RECIPE →
page 175

<IN THE SOOP BTS ver. Season 2> EP. 3

Jung Kook's toast is finished! 정국표 토스트 완성!

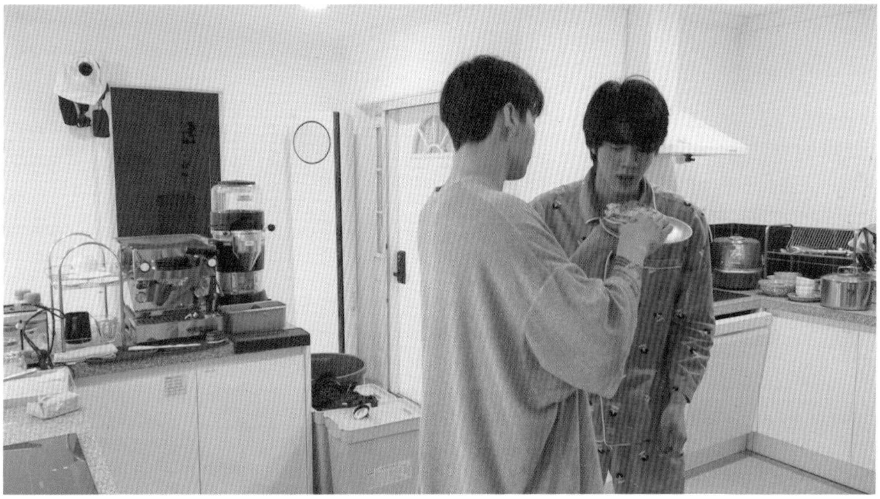

Jin tries Jung Kook's toast 정국이 만든 토스트를 먹어보는 진

Jung Kook makes some toast　　토스트를 만드는 정국

Jung Kook

*Crispy on the outside,
moist on the inside.*

겉은 바삭하게, 속은 촉촉하게.

<IN THE SOOP BTS ver. Season 2> EP. 3

Jung Kook is not the only member who enjoys toast for breakfast. j-hope has his own toast recipe, too. In a vlog filmed in the U.S. when he went there to shoot his music video for "Chicken Noodle Soup (Feat. Becky G)," j-hope shared his own recipe for "Hui-mang (Hope) Toast." It featured a simple combination of hotel breakfast food such as toast, jam, scrambled eggs, and bacon.

Toast is a popular choice for a light breakfast or a simple picnic. Jung Kook and j-hope's quick and easy toast recipes will not only fill you up but also give you the energy you need for a wonderful day.

아침 식사로 토스트를 즐기는 멤버는 정국뿐만이 아니다. 제이홉도 자신만의 토스트 레시피를 가지고 있다. 제이홉은 'Chicken Noodle Soup (Feat. Becky G)'의 뮤직비디오 촬영을 위해 미국에 갔을 때 촬영한 브이로그에서 자신의 '희망 토스트' 레시피를 공개했다. 이 토스트는 식빵과 잼, 스크램블드에그, 베이컨 등 호텔 조식 메뉴들을 간단하게 조합해 만들 수 있다는 것이 특징이다.

토스트는 가벼운 아침 메뉴 혹은 간편하게 준비하는 피크닉 메뉴로 인기다. 정국과 제이홉의 레시피를 따라 쉽고 빠르게 만든 토스트는 빈속을 든든하게 채워 주고 즐거운 하루를 보내기 위한 에너지를 줄 것이다.

"[VLOG] j-hope (2019.10.03)"

j-hope makes "Hui-mang (Hope) Toast" '희망 토스트'를 만드는 제이홉

RECIPE →
page 174

<IN THE SOOP BTS ver. Season 2> EP. 5

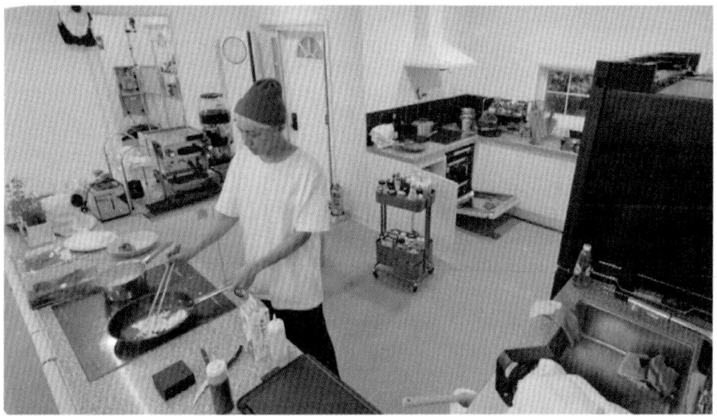

The process of making j-hope's "Hui-mang (Hope) Toast"　제이홉의 '희망 토스트' 조리 과정

BTS Choice

Q.

Do you love or hate raisins in bread?

식빵에 건포도 들어간 것을 좋아한다 vs 싫어한다?

<Run BTS!> EP. 132

V

I like it. 나는 좋아하는데.

RM

I actually like it. 나는 좋아해 사실.

Jung Kook

I do eat it. Bread with raisins—it's delicious!
나도 먹긴 하는데. 건포도 들어간 식빵, 맛있습니다!

SUGA

Aren't raisins savory? 건포도 고소한데?

j-hope

I really hate it. I can't stand them (raisins).
저는 진짜 극 불호입니다. 저는 (건포도를) 참을 수 없어요.

Jin

Why would you put raisins in bread?
I don't understand.
식빵에 건포도를 왜 넣어 먹습니까? 저는 이해가 안 돼요.

Jimin

Bread is already perfect as it is.
식빵은 식빵만으로도 완벽합니다.

A Pinch *of* Korean

겉바속촉

geot-ba-sok-chok

The phrase "crispy on the outside and moist on the inside," said by Jung Kook as he made toast, can be shortened into four characters in Korean: "geot-ba-sok-chok." This expression is an abbreviation using four letters from the Korean sentence with the same meaning. Despite being a nonstandard word, it has been widely used in recent years. You can use this expression to describe the taste of perfectly baked bread or nicely fried food. "This is geot-ba-sok-chok!"

geot		*ba-sa-ka-da*		*sok*		*chok-cho-ka-da*
겉	+	**바삭하다**	+	**속**	+	**촉촉하다**
outside		crispy		inside		moist

3

Slurp! Noodle Dishes

후루룩 면 요리

Kalguksu 칼국수

Jjajang Ramyeon 짜장 라면

- SLURP! NOODLE DISHES -

Kalguksu

Noodle Soup

칼국수

#Full_of_Umami #Boiled #Noodles

RECIPE →

page 178

The breakfast on the second day of <IN THE SOOP BTS ver. Season 2> was *haemul kalguksu* (noodle soup with seafood) with lots of seafood. *Kalguksu* (noodle soup) is a dish made by boiling noodles in a broth with vegetables, seafood, or meat. To make *haemul kalguksu* with lots of clams, you must put a lot of work into preparing the ingredients. While Jin prepared the broth and noodles, SUGA helped out by chopping the vegetables, and Jung Kook joined in by preparing the clams. Their teamwork shone through as they remained focused on their tasks even while they had a great time chatting with each other. The result of their fantastic teamwork was *haemul kalguksu* with a very rich clam broth! With the addition of j-hope, who was the last to arrive in the kitchen that morning, a four-member *haemul kalguksu* group was formed. (The other three members chose sleep over *kalguksu.*) Jin, SUGA, j-hope, and Jung Kook enjoyed a leisurely breakfast as they chatted with each other.

<인더숲 BTS편 시즌 2>에서의 둘째 날 아침 메뉴는 해물을 잔뜩 넣은 해물 칼국수였다. 칼국수는 국물과 면에 채소, 해물, 고기 등을 넣고 끓여서 만드는 요리다. 그중에서도 조개를 잔뜩 넣은 해물 칼국수는 재료 손질에 손이 많이 가는데, 진이 국물과 면을 준비하는 동안 슈가와 정국이 자연스럽게 합류하여 각각 채소를 썰고 조개를 손질하며 일을 나누었다. 즐겁게 대화를 주고받으면서도 각자 맡은 일에 집중하는 모습에서 그들의 팀워크가 엿보였다. 아름다운 팀워크의 결과로 완성된 것은 바로 진한 조개 국물의 해물 칼국수! 이날 아침에는 가장 늦게 부엌에 온 제이홉까지 더해 네 명의 해물 칼국수 모임이 결성되었다. (나머지 세 명은 칼국수 대신 잠을 택했다.) 진과 슈가, 제이홉, 정국은 칼국수를 먹으면서도 도란도란 이야기를 나누며 여유로운 아침 식사를 즐겼다.

<IN THE SOOP BTS ver. Season 2> EP. 3

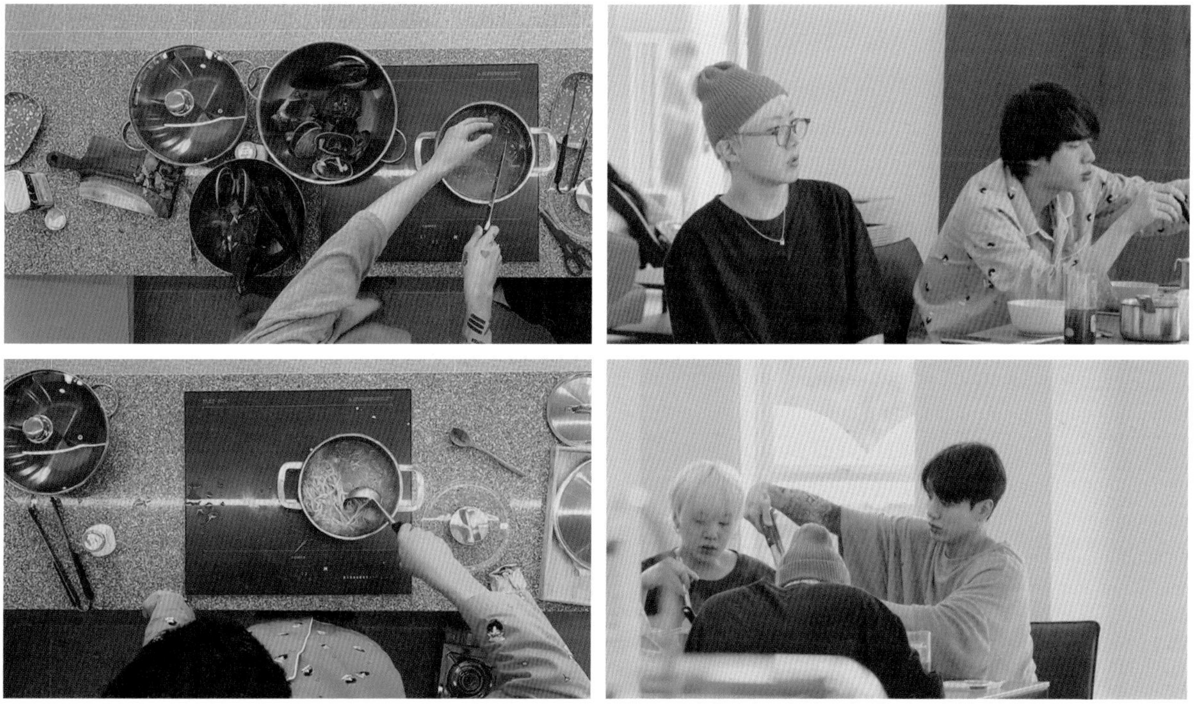

The *haemul kalguksu* group is formed!　해물 칼국수 모임 결성!

j-hope

Let me clean up.
You guys can all go.

제가 치울게요. 다들 가세요.

Jin

No, let's do it together.

아니야, 같이 치워.

SUGA

Really, j-hope,
thanks so much.

진짜, 홉아. 고맙다.

Jung Kook

I shouldn't be saying this,
being the maknae and all,
but... thanks a million!

아, 막내로서 이러면 안 되는데, 진짜 너무... 고맙다!

<IN THE SOOP BTS ver. Season 2> EP. 3

Jin, who loves noodles, made a noodle dish again in an episode of <Jin's Traditional Alcohol Journey>*. One cold, windy morning, Jin's hot *sagol kalguksu* (beef bone noodle soup) warmed everyone up. Jin cooked the *kalguksu* in a thick broth, which is his favorite style, and it tasted so good that he received compliments from a culinary expert. Once you have the noodles prepared, you can make many different types of *kalguksu* depending on the broth you use. Do you want to try your own hand at cooking a noodle dish loved by Jin? If so, try making some *kalguksu*.

면을 사랑하는 진은 *<취중진담>에서도 면 요리를 만들었다. 찬바람이 부는 어느 아침, 진이 끓인 뜨끈한 사골 칼국수는 몸을 훈훈하게 덥혀 주었다. 진은 자신이 좋아하는 스타일로 걸쭉한 국물의 칼국수를 만들었는데, 요리 전문가에게도 칭찬 받을 만큼 맛이 훌륭했다. 칼국수는 면만 미리 준비해 두면 어떤 국물을 내느냐에 따라 해물부터 사골까지 다양한 맛으로 만들어 먹을 수 있다. 진이 사랑하는 면 요리에 도전해 보고 싶은가? 그렇다면 칼국수를 만들어 보자.

* <Jin's Traditional Alcohol Journey> is BTS' original four-episode video series starring Jin and culinary researcher Baek Jong-won, where Jin learns how to make traditional Korean liquor.

<취중진담>은 진과 요리연구가 백종원이 함께 출연한 방탄소년단 자체 콘텐츠로, 진이 한국의 전통주 만드는 법을 배우는 이야기를 담은 4편 분량의 영상 시리즈이다.

RECIPE →

page 178

<Jin's Traditional Alcohol Journey> EP. 4

Jin makes *kalguksu* and bulgogi　　칼국수와 불고기를 만드는 진

Jin

I do love noodles.

또 제가 면 러버(lover) 아니겠습니까?

<Jin's Traditional Alcohol Journey> EP. 4

BTS *Choice*

Q.

Which is your preferred broth style for *kalguksu*?

선호하는 칼국수 국물 스타일은?

<IN THE SOOP BTS ver. Season 2> EP. 3

Jin, SUGA

Thick broth 걸쭉한 국물

Jung Kook

Clear broth 깔끔한 국물

Q.

What do you crave when you are on a diet?

다이어트 중 먹고 싶은 요리는?

"[EPISODE] 지민 (Jimin) 'Set Me Free Pt.2' MV Shoot Sketch - BTS (방탄소년단)"

Jimin

Delicious kimchi with *kalguksu*
맛있는 김치랑 칼국수

A Pinch *of* Korean

칼국수 + 수제비

Kalguksu* and *Sujebi

<Jin's Traditional Alcohol Journey> EP. 4

<Run BTS!> EP. 142

In episode 142 of <Run BTS!>, BTS switched from making *kalguksu* to *sujebi* (수제비; hand-pulled dough soup). The ingredients and recipe for *sujebi* is similar to those of *kalguksu*, but instead of noodles, *sujebi* contains bite-sized pieces of dough that are hand-pulled from thinly rolled out flour dough. *Sujebi* is the name of both the hand-pulled dough pieces and the dish itself. If you add knife-cut noodles to the broth, it is called *kalguksu* (칼국수); if you add thin hand-pulled dough pieces, it is called *sujebi* (수제비); and if you add both, it is called a combination of the two names, *kaljebi* (칼제비).

- SLURP! NOODLE DISHES -

Jjajang Ramyeon

Instant Noodles in Black Bean Sauce

짜장 라면

#Sweet&Savory #Boiled #Noodles

RECIPE →

page 182

Jjajang ramyeon (instant noodles in black bean sauce) appears frequently in BTS' original content. On <BTS BON VOYAGE Season 4>, V brought some instant *jjajang ramyeon* all the way from Korea to New Zealand. j-hope and Jung Kook created an incredible dish by adding onions and beef to the *jjajang ramyeon*. That day, the *jjajang ramyeon* was not cooked in the usual way but was made according to Jung Kook's special recipe. Jung Kook's *jjajang ramyeon* received much praise from the other members. Even SUGA, who said he was not hungry, ended up picking up his chopsticks at the sight of everyone enjoying the dish. For lunch, BTS devoured seven servings of *jjajang ramyeon* and three servings of instant rice, and was able to replenish their energy for their afternoon activities.

짜장 라면은 방탄소년단의 다양한 자체 콘텐츠에서 자주 등장한다. <BTS 본보야지 시즌 4>에서 뷔는 인스턴트 짜장 라면을 한국에서 뉴질랜드까지 챙겨 갔는데, 제이홉과 정국은 뷔가 가져온 짜장 라면에 양파와 소고기를 더해 훌륭하게 요리해냈다. 이날 짜장 라면은 보통의 조리법과 다르게, 정국의 특별 레시피를 따라 만들어졌다. 멤버들은 정국의 짜장 라면에 칭찬을 아끼지 않았는데, 배가 고프지 않다고 했던 슈가도 모두가 맛있게 먹는 모습을 보고 결국 젓가락을 들고 말았다. 그렇게 방탄소년단은 점심으로 짜장 라면 7인분과 즉석밥 3인분을 해치우며 오후 활동을 위한 에너지를 충전했다.

<BTS BON VOYAGE Season 4> EP. 7

j-hope and Jung Kook make *jjajang ramyeon* while Jimin watches 짜장 라면을 만드는 제이홉과 정국, 그리고 둘이 요리하는 모습을 지켜보는 지민

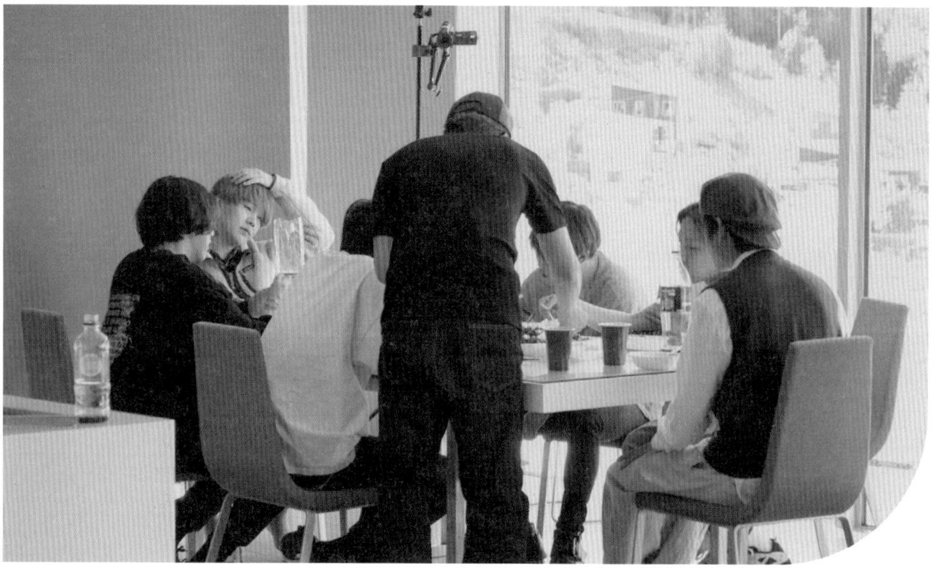

SUGA

You guys eat really well.
I think our team eats really well.

너네 진짜 잘 먹는다. 우리 팀 진짜 잘 먹는 것 같아.

<BTS BON VOYAGE Season 4> EP. 7

<Run BTS!> EP. 151

BTS orders *galbitsal chapaguri* unanimously! 방탄소년단 전원 갈빗살 짜파구리 주문!

Jjajang ramyeon also appears on <Run BTS!>. In episode 151 of <Run BTS!>, BTS was assigned individual rooms at a hotel and enjoyed a relaxing hotel staycation. A variety of food and activities were available as room service, but there was one dish that everyone ordered unanimously. That dish was "*galbitsal* (ribeye) *chapaguri**." Also known as "ram-don" because of a Korean movie, *chapaguri* is made by combining two types of *ramyeon* (instant noodles)–*jjajang ramyeon* and regular *ramyeon* with thick noodles–and topping them off with grilled beef.

The *jjajang* of *jjajang ramyeon* is a sauce that is made with Korean-style *chunjang* (black bean paste) stir-fried with pork and onions. It is known for its rich and savory flavor. On top of the *jjajang ramyeon* flavor, *chapaguri* has an added spiciness that comes from the regular *ramyeon*, so you can enjoy the spicy-savory sauce, along with the different noodle textures.

짜장 라면은 <달려라 방탄>에서도 등장한다. <달려라 방탄> 151화에서 방탄소년단은 각자 방 하나씩을 배정받고 여유로운 호캉스를 즐겼다. 다양한 먹거리와 놀거리를 룸서비스로 주문할 수 있었던 이곳에서 모든 멤버가 주문한 메뉴가 하나 있었으니, 바로 '갈빗살 짜파구리'였다. 한국 영화를 통해 'ram-don'이라는 이름으로도 알려진 짜파구리는 짜장 라면(짜파게티)과 면이 굵은 일반 라면(너구리) 두 가지를 섞고, 구운 소고기를 토핑하여 만든다.

짜장 라면에서 짜장은 콩으로 만든 한국식 춘장을 돼지고기와 양파 등에 볶아 만든 소스로, 고소하고 감칠맛이 나는 것이 특징이다. 짜파구리는 이러한 짜장 라면에 일반 라면의 매콤한 맛이 더해져 매콤짭짤하고, 면의 식감 또한 다양하다.

---

* *Chapaguri* is a combination of the names of two types of instant noodles produced by a Korean company called Nongshim. *Chapa* comes from *Chapaghetti*, the name of a *jjajang ramyeon*, while *guri* comes from *Neoguri*, the name of a regular ramyeon with thicker noodles. These two products are the ones typically used to make *chapaguri*.

RM

Why is chapaguri so good?

짜파구리는 왜 이렇게 맛있는 거지?

SUGA

It's a combination that can't go wrong.

맛없을 수 없는 조합이지.

<IN THE SOOP BTS ver. Season 1> EP. 1

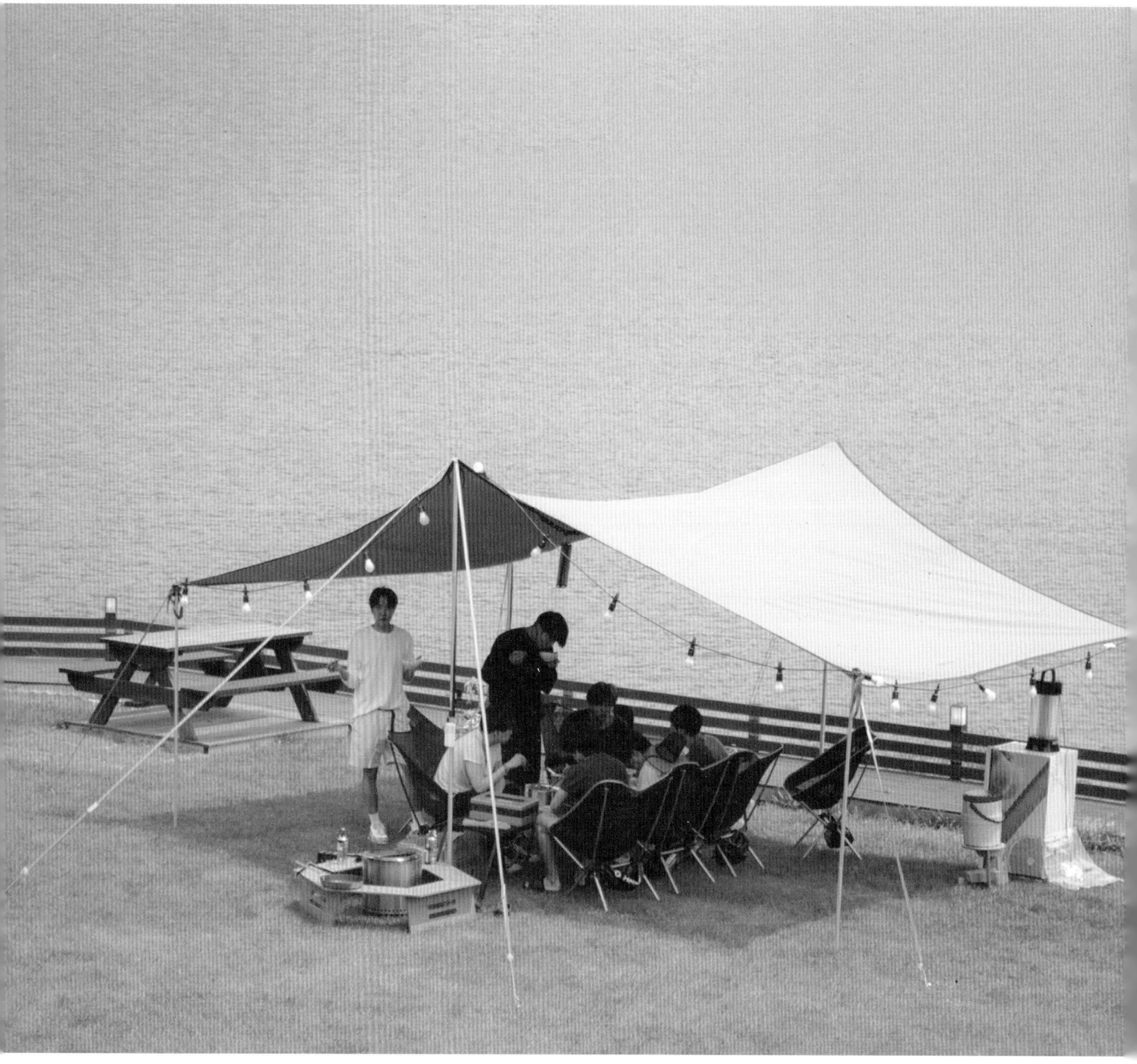

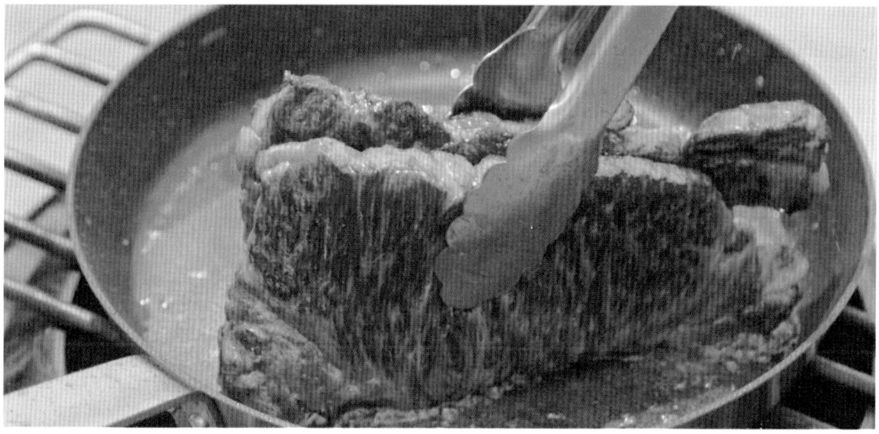

BTS enjoys *chapaguri* together 함께 짜파구리를 먹는 방탄소년단

<BTS BON VOYAGE Season 4> EP. 7

Jung Kook's hearty *jjajang ramyeon* 든든한 정국표 짜장 라면

Just by adding a few ingredients to your instant *jjajang ramyeon*, you can create a more flavorful and nutritious meal. If you are looking for a quick and hearty meal, try making Jung Kook's *jjajang ramyeon*.

인스턴트 짜장 라면에 몇 가지 재료만 추가하면 맛과 영양이 더해진 한 끼 식사를 만들 수 있다. 간편하면서도 든든한 식사 메뉴를 찾고 있다면, 정국표 짜장 라면을 만들어 보자.

RECIPE →

page 182

BTS *Choice*

Q.

What dish would you like to order from room service at a hotel?

호텔에서 룸서비스로 식사를 주문한다면 시킬 메뉴는?

<Run BTS!> EP. 151

BTS

Galbitsal chapaguri 갈빗살 짜파구리

A Pinch *of* Korean

짜장?

or

자장?

Jjajang? Jajang?

Jjajang (짜장) of *jjajang ramyeon* (짜장 라면) was originally called *jajang* (자장); but as more people started to pronounce it *jjajang*, *jjajang* also became a standard word. Typical dishes that use *jjajang* sauce include *jjajang ramyeon* (짜장 라면), *jjajangmyeon* (짜장면), which is a dish where the sauce is mixed with thick noodles, and *jjajang tteokbokki* (짜장 떡볶이), which is stir-fried rice cakes with *jjajang* sauce.

4

Spicy Meat Dishes

매콤한 고기 요리

Dakgalbi 닭갈비

Jeyuk Bokkeum 제육볶음

- SPICY MEAT DISHES -

Dakgalbi

Spicy Stir-Fried Chicken

닭갈비

#Spicy #Stir-Fried #Chicken

RECIPE →

page 186

The last meal on <IN THE SOOP BTS ver. Season 2> was spicy *dakgalbi* (spicy stir-fried chicken). *Dakgalbi* is a popular Korean dish where chicken, cabbage, sweet potato, and perilla leaves are stir-fried in a spicy *gochujang* (Korean red chili paste)-based sauce. Jin and SUGA carefully stir-fried the piles of vegetables and chicken, making sure they didn't overflow from the square pans.

Jin's helpful tip is to cook the vegetables first by placing them on the bottom of the square pans. This causes the volume of the vegetables to decrease, making it easier to stir-fry everything together. Once the *dakgalbi* was done, BTS gathered around the table one by one and enjoyed their last meal on <IN THE SOOP BTS ver. Season 2>.

<인더숲 BTS편 시즌 2>에서의 마지막 식사는 매콤한 닭갈비였다. 닭갈비는 닭고기, 양배추, 고구마, 깻잎을 매콤한 고추장 양념에 볶아 먹는 한국의 대중적인 요리다. 진과 슈가는 팬에 쌓인 채소와 고기가 팬 밖으로 넘쳐 떨어지지 않도록 조심스럽게 닭갈비를 볶았다.

진의 팁처럼 채소를 팬 아래쪽에 놓아 숨을 먼저 죽인 후에 볶으면 채소의 부피가 작아져 볶기가 더 수월하다. 닭갈비가 완성되자 차례대로 식탁으로 모인 방탄소년단은 <인더숲 BTS편 시즌 2>에서의 마지막 식사를 함께했다.

<IN THE SOOP BTS ver. Season 2> EP. 5

Jin and SUGA stir-fry *dakgalbi* 닭갈비를 볶는 진과 슈가

SUGA

Why don't we open
a dakgalbi restaurant?

닭갈비집 하나 할래요 우리?

Jin

Will you be the one stir-frying it?

네가 볶아 주는 거니?

<IN THE SOOP BTS ver. Season 2> EP. 5

Chewy glass noodles or firmly cooked *ramyeon* noodles (instant noodles) can be added to *dakgalbi*, and additional ingredients like cheese and rice cakes also go well with it. When you have almost finished the *dakgalbi*, stir-frying some rice into the leftover sauce is a must!

Jin, who saw how well Jimin was stir-frying the rice, offered him a job at the *dakgalbi* restaurant that he had just planned to open with SUGA. Their business plans got more refined as they even came up with the restaurant's name: Dak-gal-BTS. It was decided that RM would clean the grills, Jin would organize the customers' shoes, SUGA would be the cashier, j-hope would manage publicity, Jimin would stir-fry the rice, V would serve the dishes, and Jung Kook would prepare the chicken.

닭갈비에는 쫄깃한 당면이나 꼬들꼬들한 라면 사리를 추가해 먹을 수도 있고, 치즈나 떡 등의 토핑을 추가해도 잘 어울린다. 닭갈비를 거의 다 먹어 갈 즈음에는 남은 양념에 밥을 볶아 먹는 것이 필수!

야무지게 밥을 볶는 지민을 눈여겨본 진은 자신과 슈가가 닭갈비집을 차리면 그곳에서 일해 보지 않겠냐며 스카우트 제의를 했다. 식당 이름까지 '닭갈BTS'로 정해지면서 사업 계획이 고도화되었는데, RM이 불판 닦기, 진이 신발 정리, 슈가가 계산, 제이홉이 홍보, 지민이 볶음밥 볶기, 뷔가 서빙, 정국이 닭 손질을 담당하는 것으로 결론이 났다.

Jimin mixes some rice with the *dakgalbi* sauce 닭갈비 양념에 밥을 볶는 지민

RM

What is the name of the dakgalbi restaurant?

닭갈비집 이름이 뭐예요?

RM

How about "Dak-gal-BTS"?

'닭갈BTS' 어때요?

Jin

We have to decide on that now.

이제 정해야지.

V

"Dak-gal-BTS" sounds great!

'닭갈BTS' 좋은데?

<IN THE SOOP BTS ver. Season 2> EP. 5

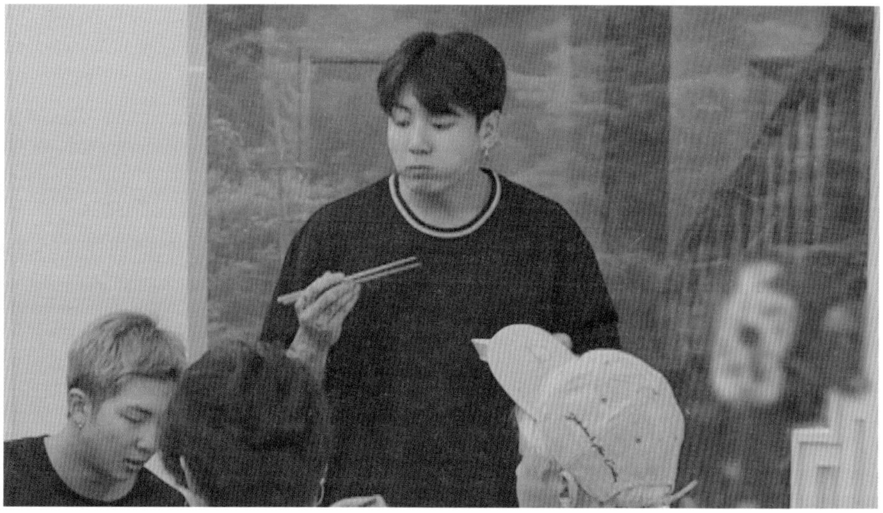

BTS sits at the table to enjoy *dakgalbi* 식탁에 모여 앉아 닭갈비를 먹는 방탄소년단

<IN THE SOOP BTS ver. Season 2> EP. 5

BTS enjoys spicy *dakgalbi* 매콤한 닭갈비를 맛있게 먹는 방탄소년단

When you're craving something spicy, how about *dakgalbi* and fried rice? The spiciness of the *dakgalbi* will wash away your stress, and the fried rice at the end will soothe your stomach, leaving you feeling full and content.

매콤한 요리를 먹고 싶은 날, 닭갈비와 볶음밥은 어떨까? 입안 가득 퍼지는 닭갈비의 매콤함이 스트레스를 날려 주고, 마무리로 먹는 볶음밥은 매운 속을 달래 주며 행복한 포만감까지 선사할 것이다.

RECIPE →
page 186

A Pinch *of* Korean

Korean Expressions of Spiciness

In Korea, where spicy foods are common, there are many adjectives describing spiciness. These words are then further subdivided into many different types of spiciness. Let's take a look at how the expressions of spiciness differ below:

Expressions of Spiciness	Meaning	Related Foods
mae-kom-ha-da **매콤하다**	spiciness felt in the mouth	*dakgalbi, jeyuk bokkeum* (spicy stir-fried pork)
al-ssa-ha-da **알싸하다**	spiciness that hits the nose or mouth very sharply	raw garlic
kal-kal-ha-da **칼칼하다**	spiciness felt in the throat	*kimchi jjigae* (kimchi stew)

Jeyuk Bokkeum

Spicy Stir-Fried Pork

제육볶음

RECIPE →

page 190

V and Jung Kook made *jeyuk bokkeum* (spicy stir-fried pork) and *gyeran mari* (rolled omelet) under Jin's directions on the "Avatar Chef King" episode of <Run BTS!>. In the beginning, they looked a little bit lost, as they didn't have a clear understanding of the rules. However, over time, they demonstrated excellent teamwork and were able to finish the dishes in time.

뷔와 정국은 <달려라 방탄> '아바타 요리왕' 편에서 진의 지령을 들으며 제육볶음과 계란말이를 만들었다. 둘은 처음에 규칙을 잘 이해하지 못하고 우왕좌왕했지만, 시간이 갈수록 손발을 척척 맞추어 제한 시간 안에 요리를 완성해냈다.

<Run BTS!> EP. 102

V and Jung Kook show off the best teamwork
최고의 팀워크를 발휘하는 뷔와 정국

Jimin

V said he cooks these days.

뷔가 요즘에 요리한다고 했었는데?

RM

He will show us his cooking skills any minute now.

뷔가 요리 실력 보여 줄 거야, 이제 나온다.

<Run BTS!> EP. 102

And of course, the results were a success. In particular, the *jeyuk bokkeum* received an enthusiastic response. This was probably because it was made with pork belly, BTS' favorite. The other members came out to taste the finished dish and couldn't stop eating it. When V, who said that he had been cooking a lot recently, received a lot of compliments on his dish, he said with confidence:

물론 결과물도 성공적이었다. 멤버 모두가 좋아하는 삼겹살을 사용하여 만들었기 때문인지, 특히 제육볶음에 대한 반응이 뜨거웠다. 완성된 요리를 맛보러 나온 멤버들은 계속 젓가락이 가는 제육볶음을 맛있게 먹었다. 최근에 요리를 자주 했다는 뷔는 자신이 만든 요리가 칭찬을 받자 자신감 넘치는 모습으로 이렇게 말했다.

V

Hard work never betrays you.

노력은 결코 배신하지 않더라고요.

<Run BTS!> EP. 102

V uses his fingers to estimate the length of a piece of large green onion
손가락으로 대파 한 도막의 길이를 가늠해 보는 뷔

Jeyuk bokkeum and *gyeran mari* made by V and Jung Kook　뷔와 정국이 완성한 제육볶음과 계란말이

Jin gives V and Jung Kook orders with a walkie-talkie 무전기로 뷔와 정국에게 지령을 내리는 진

Deliciously cooked *jeyuk bokkeum* 맛있게 익어 가는 제육볶음

Jeyuk bokkeum is a meat dish that BTS, who loves meat, has enjoyed ever since they were trainees. The spicy and oily flavor of the pork marinated in a *gochujang* (Korean red chili paste)-based sauce pairs well with rice, which makes it a common side dish in Korea. If you have some leafy vegetables, it is recommended that you put the rice and meat on top and eat it as a wrap. The flavor of fresh vegetables will neutralize the greasy taste and make a more nutritionally balanced meal.

제육볶음은 고기를 좋아하는 방탄소년단이 연습생 때부터 즐겨 먹었던 고기 요리다. 고추장 양념이 밴 고기의 매콤하고 기름진 맛이 밥과 잘 어울려서 한국에서는 밥반찬으로도 흔하게 먹는다. 쌈을 해 먹을 만한 채소가 있다면 밥과 고기를 함께 올려 쌈을 싸 먹는 것을 추천한다. 신선한 채소의 맛이 느끼함을 잡아 주고, 영양의 균형도 맞춰 줄 것이다.

RECIPE →
page 190

As the judge, j-hope tastes the dishes made by V and Jung Kook　뷔와 정국이 만든 요리를 시식하는 심사위원 제이홉

<Run BTS!> EP. 102

V and Jung Kook make the seasoning 양념을 만드는 뷔와 정국

BTS *Choice*

Q.

Which dish did BTS enjoy often as trainees?

방탄소년단이 연습생 때 자주 먹었던 요리는?

BTS

Jeyuk bokkeum 제육볶음

A Pinch *of* Korean

구(9)+이(2)

Describing "Meat for Grilling" Using Body Language

In the "Avatar Chef King" episode of <Run BTS!>, the avatar chefs were not allowed to ask or answer any questions and were only allowed to listen to the member who was giving instructions. Under these rules, Jin said to V, "Describe the type of meat Jung Kook brought using body language." Hearing this, j-hope asked, "How do you describe whether the meat is for stewing or grilling using body language?" Jin jumped up and made some shapes with his body while shouting, "Gu-i!" (구이!; Grilled meat!) The shapes that Jin made were the numbers 9 and 2. When talking about meat for grilling, the pronunciation of gu-i (구이; grilled meat) is the same as that of the numbers 9 (구; gu) and 2 (이; i) in Korean, which is why he described it this way.

5

Dishes for Special Feasts

특별한 만찬

Miyeokguk 미역국
Japchae 잡채
Tteokguk 떡국
Bulgogi 불고기

- DISHES FOR SPECIAL FEASTS -

Miyeokguk

Seaweed Soup

미역국

RECIPE →

page 194

After the "BTS <Yet To Come> in BUSAN" concert in 2022, j-hope, Jimin, V, and Jung Kook started a live broadcast. The broadcast took place in a very relaxed setting where the four casually talked about the concert. Jimin, whose birthday was October 13, just two days before the concert, said he came to Busan a few days early to visit his parents and enjoyed some of their *miyeokguk* (seaweed soup). Jung Kook made everyone laugh by saying that he also had *miyeokguk* at his parents' house to celebrate Jimin's birthday.

2022년 'BTS <Yet To Come> in BUSAN' 콘서트 후 제이홉, 지민, 뷔, 정국은 라이브 방송을 켰다. 넷은 편안한 분위기 속에서 콘서트 회포를 풀었는데, 콘서트 이틀 전인 10월 13일이 생일이었던 지민은 부산의 부모님 댁에 미리 내려와 부모님이 해 주신 미역국을 먹었다고 했다. 정국도 지민의 생일을 맞이해 자신의 부모님 댁에서 미역국을 먹었다고 말해 모두를 웃게 만들었다.

Jimin

Why did you eat miyeokguk?

넌 왜 미역국 먹었어?

Jung Kook

Because it was your birthday!

지민이 형 생일이어서!

Weverse (2022.10.15)

Weverse (2022.10.15)

j-hope, Jimin, Jung Kook, and V talk together on the live broadcast 라이브 방송에서 이야기를 나누는 제이홉, 지민, 정국과 뷔

RM

(On a live broadcast the day after his birthday)
On my birthday,
I had miyeokguk with my family.
To be honest, I shouldn't have been the one eating it.
I should have made it for my mother.
I just ate it because my mother
wanted to make it for me.

... I had tasty galbi (ribs), miyeokguk, and
japchae (stir-fried glass noodles and vegetables),
and I went hiking, and yeah, that's about it.

(자신의 생일 다음 날 진행한 라이브 방송에서)
생일 때 가족들이랑 미역국 먹고,
사실 미역국은 제가 먹을 게 아니고
저희 어머니께 제가 끓여 드려야 맞는데,
어머니께서 미역국을 끓여 주고 싶다고 하셔서 얻어먹은 거죠.

... 갈비, 미역국, 잡채 이렇게 맛있게 먹었고,
등산하고, 뭐 그랬어요.

Weverse (2021. 09. 13)

In Korea, there is a custom of eating *miyeokguk* on your birthday every year, so it is the first dish that comes to mind when you think of birthdays. It is also one of the most common postpartum care dishes, as it is known to help women who have given birth recover quickly. In fact, this is the reason why it has become a symbol of birthdays. *Miyeokguk* is characterized by the seaweed's firm texture and the salty, ocean-scented broth.

Due to the slippery texture of the seaweed, there is a superstitious belief that eating *miyeokguk* on the day of a test causes students to "slip" and fail, so the dish is avoided on test days.

한국에서는 매년 생일날 미역국을 챙겨 먹는 풍습이 있어서, 미역국은 생일 하면 가장 먼저 떠오르는 요리다. 미역국은 산모들이 빨리 회복할 수 있도록 도와주는 요리로 알려져 있어서 산후조리를 할 때 많이 먹는데, 자연스럽게 생일을 상징하는 대표적인 요리가 되었다. 미역국은 꼬들꼬들한 미역의 식감과 바다향이 나는 짭조름한 국물 맛이 매력적이다.

미끈거리는 미역의 식감 때문에 시험 날 먹으면 '미끄러진다'라는 미신이 있어 중요한 시험이 있는 날에는 잘 먹지 않는다는 점도 재미있다.

"[2020 FESTA] BTS (방탄소년단) '방탄생파' #2020BTSFESTA"

Jin soaks seaweed in water　미역을 물에 불리는 진

Jin serves *miyeokguk* to the members　멤버들에게 미역국을 덜어 주는 진

Jimin brings the cake that he decorated with SUGA　슈가와 함께 꾸민 케이크를 가지고 오는 지민

Jimin lights the candles on the cake　케이크 초에 불을 붙이는 지민

Depending on the ingredients used to make the broth, the flavor of *miyeokguk* varies slightly. Though the beef *miyeokguk* that Jin made during the 2020 FESTA in BTS' birthday party video is a common type, depending on your preference, you can also make a savory broth with seafood or a clear vegan broth with golden oak mushrooms.

미역국의 맛은 어떤 재료로 국물을 내는지에 따라 조금씩 달라진다. 2020 FESTA의 '방탄생파' 영상에서 진이 만든 소고기 미역국도 흔히 먹지만, 취향에 따라 해산물로 시원한 국물을, 표고버섯으로 맑은 비건 국물을 낼 수도 있다.

RECIPE →
page 194

"[2020 FESTA] BTS (방탄소년단) '방탄생파' #2020BTSFESTA"

j-hope decorates the room for the birthday party 생일파티 장소를 꾸미는 제이홉

BTS Choice

Q.

Which dish would you recommend to ARMY who wants to make Korean food at home?

집에서 한식을 만들어 먹고 싶은 아미에게 추천해 주고 싶은 요리는?

<BTS RECIPE BOOK>

j-hope

Miyeokguk 미역국

A Pinch *of* Korean

HAPPY BIRTHDAY

생일 축하합니다!

Sing the Happy Birthday Song in Korean

ARMY and BTS sang the birthday song for Jimin at the "BTS <Yet To Come> in BUSAN" concert in 2022. For BTS' upcoming birthdays, let's practice singing the happy birthday song in Korean. You just have to put the person's name in the OOO part.

♫ *Saeng-il chu-ka-ham-ni-da.* (× 2)
생일 축하합니다.
Happy birthday to you.

♫ *Sa-rang-ha-neun* OOO, (× 2)
사랑하는 OOO,
Dear OOO,

♫ *Saeng-il chu-ka-ham-ni-da.*
생일 축하합니다.
Happy birthday to you.

"[EPISODE] BTS (방탄소년단) @ 'Yet To Come' in BUSAN"

- DISHES FOR SPECIAL FEASTS -

Japchae

Stir-Fried Glass Noodles and Vegetables

잡채

#Chewy #Stir-Fried #Noodles

RECIPE →

page 196

In celebration of the seventh anniversary of their debut during the 2020 FESTA, BTS reenacted their first birthday party, which they had back at the legendary "Blue House*." Just like on that day seven years ago, RM and Jin cooked the food, SUGA and Jimin decorated the cake, and j-hope, V, and Jung Kook decorated the venue. While Jin cooked *miyeokguk* (seaweed soup), RM made *japchae* (stir-fried glass noodles and vegetables).

In Korea, *japchae* is one of the most popular dishes for celebratory occasions like birthdays and holidays. *Japchae* is made of stir-fried vegetables and meat mixed with glass noodles. The flavor of the sweet soy sauce-based sauce is not strong, so it can be enjoyed by everyone. Furthermore, the harmony of crunchy vegetables, tender meat, and chewy glass noodles lets you enjoy a variety of textures.

2020 FESTA에서 방탄소년단은 7번째 생일을 맞이하여 전설의 '*파란 집'에서의 첫 생일파티를 재연했다. 7년 전 그날처럼 RM과 진은 요리를 담당했고, 슈가와 지민은 케이크를, 제이홉, 뷔, 정국은 장소를 꾸몄다. 진이 미역국을 끓이는 동안 RM은 잡채를 만들었다.

한국에서 잡채는 생일이나 명절 같은 잔칫날에 빠지지 않는 인기 메뉴다. 잡채는 볶은 채소와 고기를 당면과 버무린 요리인데, 달달한 간장 베이스 양념이 자극적이지 않아 누구나 맛있게 먹을 수 있고 아삭한 채소, 부드러운 고기, 쫄깃한 당면이 어우러져 식감 또한 다채롭다.

---

* Blue house is the nickname for BTS' first house. The inside of the house, as described in the lyrics of BTS' song "Moving On" from the album <화양연화 pt.1>, was all blue, even down to its wallpaper.

파란 집은 방탄소년단의 첫 숙소다. <화양연화 pt.1> 앨범 수록곡 '이사'의 노래 가사에 설명된 것과 같이, 벽지를 포함한 숙소 내부가 온통 파란색이어서 얻은 별명이다.

"[2020 FESTA] BTS (방탄소년단) '방탄생파' #2020BTSFESTA"

RM prepares the vegetables for *japchae* 잡채에 들어갈 채소를 준비하는 RM

Jung Kook

RM, your japchae is so good!

남준이 형, 잡채 진짜 맛있는데요?

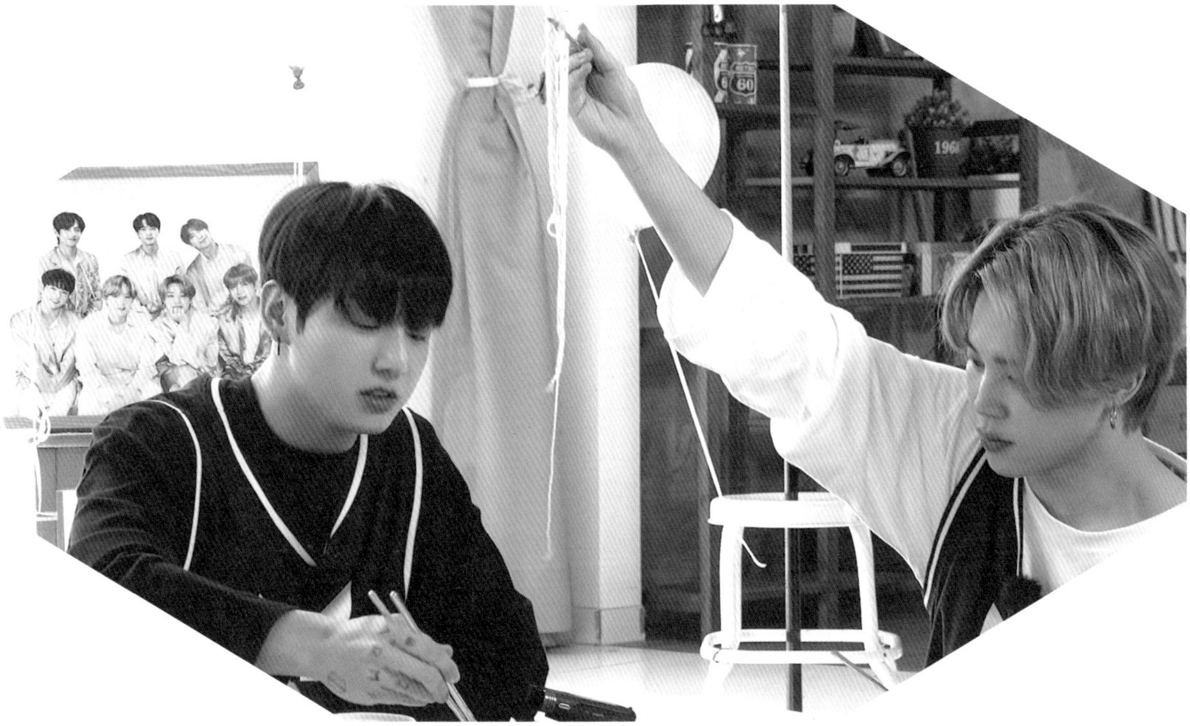

Jimin

Great job making japchae!

잡채 잘 만들었다!

"[2020 FESTA] BTS (방탄소년단) '방탄생파' #2020BTSFESTA"

After singing the happy birthday song and blowing out the candles on the cake, they were finally ready to eat. Contrary to RM's concern, his *japchae* was a great success. That day, V demonstrated the "noodle-slurping technique," which is when you slurp the noodles into your mouth in one go without biting them off and then chew the noodles in your mouth. A mouthful of *japchae* will make you go "Mmm!" like BTS. If you are looking for a party dish, a plate of sweet and savory *japchae* is highly recommended.

생일 축하 노래를 부르고 케이크 초를 끈 뒤 드디어 식사가 시작됐다. RM이 잡채를 만들면서 걱정했던 것과는 달리 잡채는 매우 성공적이었다. 이날 뷔는 '면 치기 기술'을 선보였는데, 면을 끊지 않고 한 번에 흡입한 후 입안에서 씹어 먹는 기술이다. 이렇게 잡채를 한입 가득 먹으면 방탄소년단처럼 "음!" 소리가 절로 날 것이다. 다음번 파티 요리를 고민 중이라면, 달콤 짭짤한 잡채를 적극 추천한다.

RECIPE →
page 196

BTS enjoys *japchae*　　잡채를 맛있게 먹는 방탄소년단

Jimin

(decorating the cake)
Oh, I can do it well. Don't worry.

(케이크를 꾸미면서) 아, 저는 잘해요. 걱정하지 마세요.

SUGA

I'm not worried.
Why would I be?
I know you'll do great.

걱정 안 해. 걱정을 왜 해, 잘할 건데.

"[2020 FESTA] BTS (방탄소년단) '방탄생파' #2020BTSFESTA"

A Pinch *of* Korean

잡채? / 자체?

***Japchae* and Ja-che**

The Korean buzzword "geu ja-che" (그 자체; itself) means that a feature stands out so much that it looks like the very thing itself. For example, you could say that BTS' performance is "perfection geu ja-che" or their music is "awesomeness geu ja-che." However, the pronunciations of "ja-che" (자체) and "*japchae*" (잡채) are similar, so saying "*japchae*" (잡채) instead of "ja-che" (자체) became popular with the younger generation. "Awesomeness geu *japchae*," does not mean that the *japchae* is awesome, but rather that something mentioned earlier is awesome, so don't confuse these two expressions!

HAPPY BIRTHDAY

BTS

- DISHES FOR SPECIAL FEASTS -

Tteokguk

Sliced Rice Cake Soup

떡국

#Chewy #Boiled #Soup

RECIPE →

page 198

One cold day in 2018, RM and V climbed a mountain at dawn to carry out the <Run BTS!> punishment of watching the sunrise from Achasan Mountain. Jin, the Chef King of BTS, also woke up early to prepare *tteokguk* (sliced rice cake soup) as box lunches for the hiking duo.

In Korea, it is customary to enjoy *tteokguk* on Seollal, the first day of both the Solar and Lunar years. It is said that when you eat a bowl of *tteokguk*, you become a year older. To make the perfect *tteokguk* for the new year, boil chewy sliced rice cakes in beef bone broth and top it off with colorful garnish.

2018년 어느 날, RM과 뷔는 아차산에서 일출을 봐야 하는 <달려라 방탄> 벌칙을 수행하기 위해 추운 새벽에 산을 올랐다. 이날 방탄소년단의 요리왕 진도 일찍 일어나 등산 듀오를 위한 떡국 도시락을 준비했다.

한국에는 새해 첫날인 양력 1월 1일과, 음력 1월 1일에 떡국을 먹는 풍습이 있는데, 떡국을 한 그릇 먹으면 나이를 한 살 먹는다고 말하기도 한다. 담백한 사골 국물에 쫄깃한 떡국떡을 넣어 끓이고 그 위에 알록달록한 고명을 얹으면 새해맞이 떡국이 완성된다.

<Run BTS!> EP. 44

The sunrise from Achasan Mountain　아차산의 일출

RM and V arrive at Achasan Mountain by car at dawn　새벽에 차를 타고 아차산에 도착한 RM과 뷔

V

(eating garnish)

(고명 먹는 중)

Jin

Please don't eat that.
That's the garnish.

먹지 마세요, 고명이에요.

<Run BTS!> EP. 44

RM and V arrive at Achasan Mountain early in the morning　아침 일찍 아차산에 도착한 RM과 뷔

The view of Seoul from Achasan Mountain　아차산에서 내려다본 서울의 모습

Though people usually have *tteokguk* on New Year's Day, it is also a common food you can see every day. *Tteongmandutguk* (sliced rice cake and *mandu* (dumpling) soup), which is *tteokguk* with *mandu*, is also popular. In an interview following the release of the album <MAP OF THE SOUL : 7>, Jung Kook chose *tteongmandutguk* when asked what food he could eat for a month straight. If you would like a heartier meal, consider adding *mandu* to your *tteokguk* for extra flavor and nutrition.

떡국은 주로 새해 첫날에 많이 먹지만 평상시에도 흔하게 볼 수 있는 요리다. 여기에 만두를 추가한 떡만둣국도 인기가 많은데, 정국은 <MAP OF THE SOUL : 7> 앨범 발매 당시 진행한 인터뷰에서 '한 달 내내 먹을 수 있는 음식'으로 떡만둣국을 선택하기도 했다. 조금 더 든든한 식사를 하고 싶은 날에는 떡국에 만두를 추가해 더욱 풍부한 맛과 영양을 즐겨 보자.

"BTS (방탄소년단) 7-Second Interview"

Q.

What food could you eat for a month straight?

한 달 내내 먹을 수 있는 음식은?

Jung Kook

Tteongmandutguk

떡만둣국

In Korea, many people make a wish on New Year's Day as they watch the sunrise. On the next New Year's Day, make a wish as you watch the sunrise like RM and V, and enjoy some homemade *tteokguk*. In this way, you will be able to start the new year off with higher spirits!

한국에는 새해 첫날에 일출을 보면서 소원을 비는 사람이 많다. 돌아오는 새해 첫날에는 RM과 뷔처럼 일출을 보며 소원도 빌고 떡국도 만들어 먹어 보자. 새로운 한 해를 더욱 힘차게 시작할 수 있을 것이다!

RECIPE →
page 198

<Run BTS!> EP. 44

V and RM enjoy the *tteokguk* made by Jin 진이 만들어 준 떡국을 먹는 뷔와 RM

V and RM wish ARMY a happy New Year 아미에게 새해 인사를 하는 뷔와 RM

A Pinch of Korean

New Year's Greetings in Korean

"BTS (방탄소년단) 2023 Happy Seollal Greeting"

In Korea, it is customary for younger people to perform a New Year's bow for their elders as a New Year's greeting on New Year's Day. Typically, the younger people say, "Sae-hae bok ma-ni ba-deu-se-yo!" (새해 복 많이 받으세요!; Best wishes for the new year!) as they bow, and the elders offer words of wisdom along with some money. The expression "Sae-hae bok ma-ni ba-deu-se-yo" is used not only when bowing but also as a casual New Year's greeting. How about exchanging New Year's greetings in Korean next New Year's Day?

- DISHES FOR SPECIAL FEASTS -

Bulgogi

불고기

#Sweet&Salty #Simmered #Beef

RECIPE →

page 200

To shoot his music video for "Arson," j-hope traveled to Gwangyang-si, Jeollanam-do Province. After a four-hour drive from Seoul, the first thing j-hope did in the city was eat bulgogi. Bulgogi is one of the most widely known Korean dishes, consisting of thinly sliced beef that is marinated and cooked. There are several types of bulgogi based on how the meat is cooked. Gwangyang-style bulgogi involves grilling the meat to a crisp over charcoal without any broth. That day, j-hope kept nodding in satisfaction as he savored the Gwangyang-style bulgogi's sweet and salty flavors.

'방화 (Arson)'의 뮤직비디오 촬영을 위해 전라남도 광양을 찾은 제이홉. 서울에서 네 시간을 달려 도착한 광양에서 제이홉이 가장 먼저 한 일은 바로 불고기 먹기였다. 불고기는 가장 널리 알려진 한국 음식으로, 얇게 저민 소고기를 양념하여 익혀 먹는 요리이다. 불고기는 고기를 익히는 방법에 따라 몇 가지 종류로 나뉘는데, 그중 광양식 불고기는 고기를 국물 없이 숯불에 바싹 구워 먹는 것이 특징이다. 이날 제이홉도 연신 고개를 끄덕이며 달콤하고 짭짤한 광양식 불고기의 맛을 음미했다.

"[BTS VLOG] j-hope I 확신의 J VLOG"

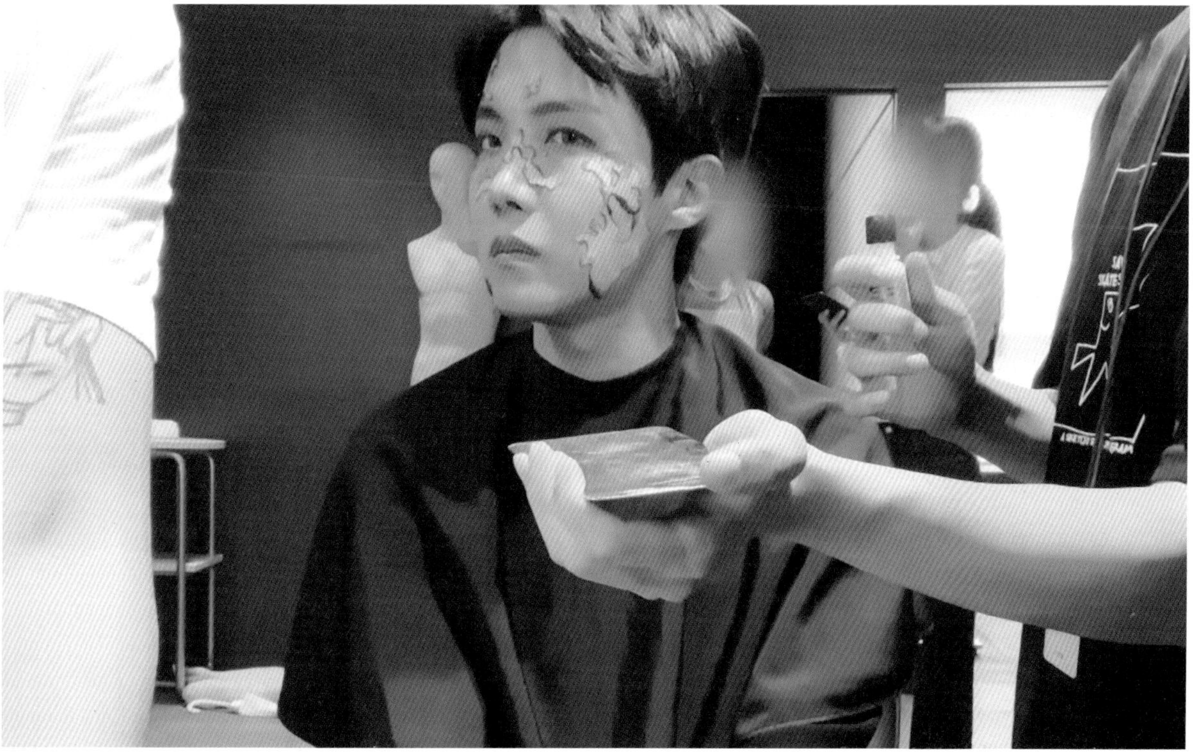

Makeup simulation for the "Arson" music video '방화 (Arson)' 뮤직비디오 메이크업 시뮬레이션

j-hope

Is this the bulgogi restaurant with the incredible bulgogi?

여기가 그렇게 맛있다는 불고기 집인가?

"[BTS VLOG] j-hope | 확신의 J VLOG"

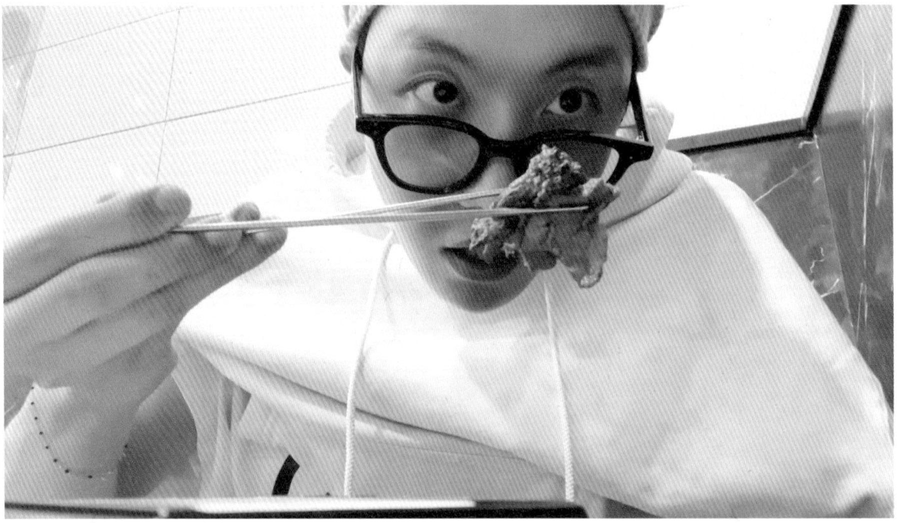

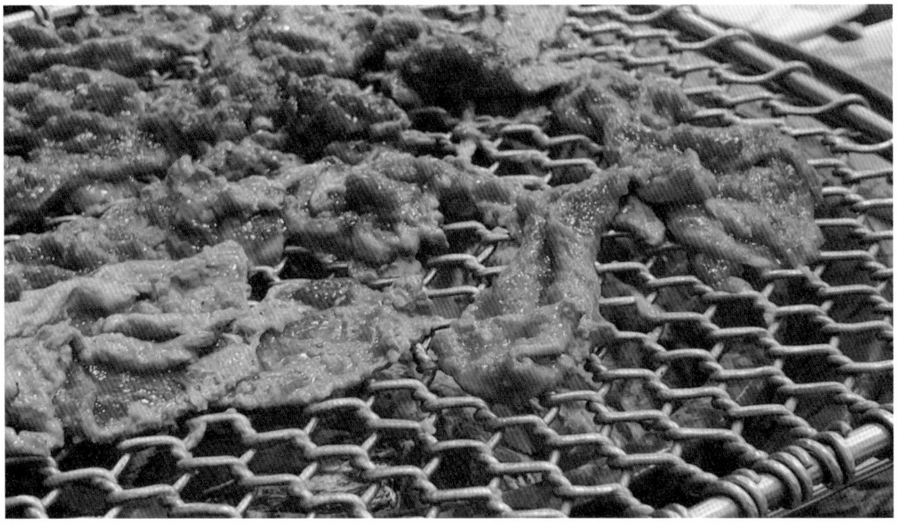

Unlike the Gwangyang-style bulgogi, the Seoul-style bulgogi is cooked by simmering meat in a broth. Jin made this Seoul-style bulgogi for breakfast one morning on <Jin's Traditional Alcohol Journey>*. He showed no signs of hesitation while making the dish. Jin simmered the marinated meat in a pot, topped it off with eggs and finely chopped scallions, and completed the steaming hot bulgogi dish in no time. A piece of well-marinated meat with a spoonful of broth can warm you up from the chilly morning air, which makes it a great breakfast dish.

반면 서울식 불고기는 광양식과 다르게 고기를 자작한 국물과 함께 끓여서 익힌다. *<취중진담>에서 맞이한 아침, 진은 아침 식사로 서울식 불고기를 만든 적이 있다. 요리를 하는 진의 손길에는 망설임이 없었다. 진이 양념에 재워둔 고기를 냄비에 담아 끓이고, 계란과 잘게 썬 쪽파를 올리자 금세 김이 모락모락 나는 불고기가 완성되었다. 간이 잘 배어든 고기 한 점을 국물과 함께 떠서 먹으면 쌀쌀한 아침 공기에 움츠러들었던 몸도 따뜻하게 데워지기에 불고기는 아침 식사 메뉴로도 좋다.

---

* <Jin's Traditional Alcohol Journey> is BTS' original four-episode video series starring Jin and culinary researcher Baek Jong-won, where Jin learns how to make traditional Korean liquor.

<취중진담>은 진과 요리연구가 백종원이 함께 출연한 방탄소년단 자체 콘텐츠로, 진이 한국의 전통주 만드는 법을 배우는 이야기를 담은 4편 분량의 영상 시리즈이다.

<Jin's Traditional Alcohol Journey> EP. 4

Jin makes bulgogi　불고기를 만드는 진

Jin

Now we just have to wait until everything starts to boil, right?

이제 끓어오르기만 기다리면 되는 거 아닙니까?

<Jin's Traditional Alcohol Journey> EP. 4

Jin enjoys the bulgogi he made　직접 만든 불고기를 먹는 진

Since it is difficult to use charcoal at home, Seoul-style bulgogi is more commonly made in Korean households. Bulgogi requires simple ingredients and is easy to make. Its fairly sweet flavor appeals to people of all ages, which is why in Korea, bulgogi is often served during holidays and celebrations when large groups of people gather.
If you ever invite a lot of guests and want to serve a main dish that everyone will love, bulgogi will be an excellent choice.

집에서는 숯불을 사용하기 어렵기 때문에 한국의 가정집에서는 흔히 서울식 불고기를 먹는다. 불고기는 재료가 단순하고 만들기도 쉽다. 게다가 달짝지근한 맛으로 남녀노소 모두의 입맛을 만족시켜, 한국에서는 여러 사람이 모이는 명절이나 잔칫날에 불고기가 자주 등장한다. 다양한 손님들을 초대한 날, 누구에게나 사랑받는 메인 요리를 내고 싶다면 불고기가 좋은 답이 되어 줄 것이다.

RECIPE ⟶
page 200

<Jin's Traditional Alcohol Journey> EP. 4

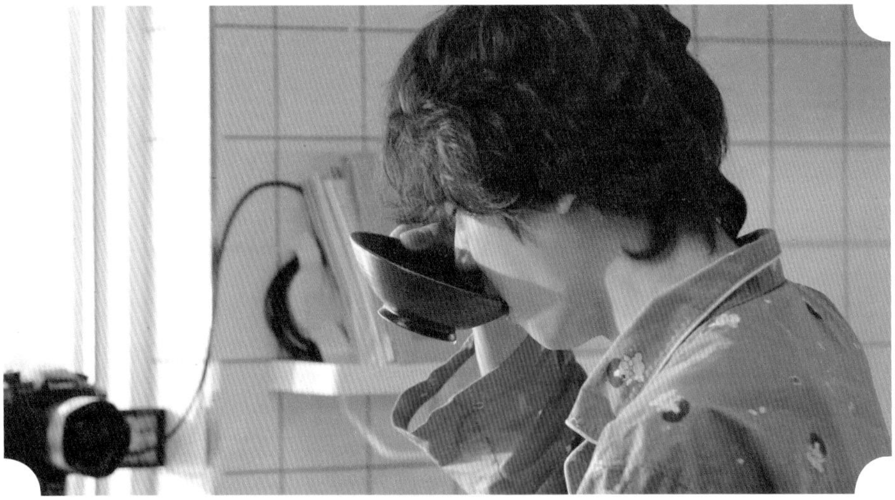

Jin has a drink while cooking 요리하면서 약주 한잔하는 진

BTS *Choice*

Q.

Which dish would you recommend to ARMY who wants to make Korean food at home?

집에서 한식을 만들어 먹고 싶은 아미에게 추천해 주고 싶은 요리는?

<BTS RECIPE BOOK>

SUGA

Bulgogi 불고기

A Pinch *of* Korean

불고기

물고기

What Happens
When Meat Meets Fire and Water?

bul **불** fire	+	*go-gi* **고기** meat	=	*bul-go-gi* **불고기** bulgogi (meat cooked over a fire)
mul **물** water	+	*go-gi* **고기** meat	=	*mul-kko-gi* **물고기** fish (meat living in water)

* Please note that "go" (고) is pronounced "kko" (꼬) in mul-kko-gi (물고기), unlike the pronunciation of bul-go-gi (불고기).

6

Perfect Pairings with Drinks

술과 어울리는 안주

Kimchijeon 김치전

Gyeran Mari 계란말이

- PERFECT PAIRINGS WITH DRINKS -

Kimchijeon

Kimchi Pancake

김치전

#Crispy #Pan-Fried #Pancake

RECIPE →

page 206

After a very busy spring and summer, <IN THE SOOP BTS ver. Season 2> was a most welcome break for BTS. One rainy day, RM reminisced about his childhood and suggested a game of *jokgu* (Korean footvolley) in the rain. Jin, Jimin, V, and Jung Kook accepted RM's offer, and the rain-soaked match was filled with unending laughter. After the game, Jin headed to the kitchen to make *kimchijeon* (kimchi pancake) to share with the other members.

봄부터 여름까지 바쁘게 달려온 방탄소년단에게 <인더숲 BTS편 시즌 2>는 선물 같은 휴식 시간이었다. 비가 오는 어느 날, RM은 어린 시절을 떠올리며 멤버들에게 빗속의 족구를 제안했다. 진, 지민, 뷔, 정국이 제안을 수락했고, 비를 흠뻑 맞으며 진행된 족구 경기에서는 웃음소리가 끊이지 않았다. 족구 후 진은 부엌으로 향했는데, 멤버들과 함께 먹을 김치전을 만들기 위해서였다.

<IN THE SOOP BTS ver. Season 2> EP. 4

RM, Jin, Jimin, V, and Jung Kook play *jokgu* in the rain　빗속에서 족구를 즐기는 RM, 진, 지민, 뷔와 정국

(BTS talking about
how they played jokgu in the rain)

(빗속에서의 족구에 대해 이야기하는 멤버들)

Jimin

It felt good to be in the rain.
I played jokgu
because I wanted to
spend time with the others.

비를 한번 맞기 시작하니까 기분이 좋아서,
멤버들이랑 놀고 싶어서 족구했습니다.

RM

It was great
because it made me
feel like a kid again.

어릴 때로 돌아간 것 같아서 좋았어요.

V

We went with the flow
and did whatever we wanted.
That's what I liked most
about this break.

몸이 가는 대로 노는 것, 그게 제일 큰 휴식이 됐던 것 같아요.

<IN THE SOOP BTS ver. Season 2> EP. 4

In Korea, the combination of crispy *jeon* (batter-fried pancakes) and *makgeolli* (unrefined rice wine) is especially popular on rainy days. There are many types of *jeon*, including *pajeon* (scallion pancakes), *yukjeon* (pan-fried battered beef), *saengseonjeon* (pan-fried battered fish fillets), and *hobakjeon* (pan-fried battered zucchini). However, one of the most beloved dishes in Korea, the birthplace of kimchi, is definitely the sweet, sour, spicy, and salty *kimchijeon*.

Although Jin prefers to use a little bit of cooking oil for a lighter flavor when pan-frying *jeon*, on this day he added a generous amount of oil to the skillet to suit the tastes of the others. As Jin was not used to frying *jeon* in so much oil, the *jeon* ripped as he was trying to flip it. But thanks to BTS' choreography leader, j-hope, who cheered him on with words of encouragement as well as the "kimchi dance," Jin gained confidence and was able to finish making delicious *kimchijeon*. The *kimchijeon* turned out so good that RM, SUGA, and j-hope all complimented his cooking skills.

한국에서는 비가 오는 날이면 바삭한 전과 막걸리 조합이 유난히 사랑받는다. 전에는 파를 주재료로 한 파전, 고기에 계란물을 입혀 부친 육전, 생선전이나 호박전 등 여러 종류가 있다. 그중에서도 김치의 본고장인 한국에서 가장 사랑받는 전 중 하나는 바로 새콤, 달콤, 매콤, 짭조름한 김치전이다.

진은 전을 구울 때 기름을 적게 넣어 담백하게 부치는 것을 선호하지만, 이날만큼은 다른 멤버들의 취향에 맞춰 기름을 넉넉히 둘렀다. 기름에 튀기듯 전을 부치는 것이 진에게는 익숙하지 않아 중간에 전이 찢어지기도 했다. 그러나 안무 팀장 제이홉이 요리 중인 진의 곁에서 다정한 말과 '김치 춤'으로 응원해 준 덕에 진은 자신감을 가지고 김치전을 맛있게 완성할 수 있었다. 완성된 김치전은 RM, 슈가, 제이홉 모두가 잘 부쳤다고 극찬을 하며 맛있게 먹었다.

Jin makes *kimchijeon*　김치전을 만드는 진

SUGA

You cooked it excellently!

잘 구웠네!

j-hope

Hyung, you cooked it so well!

형, 너무 잘 구웠는데요?

Jin

It's all thanks
to the excellent "kimchi dance"
j-hope performed for me.

제이홉이 앞에서 '김치 춤' 춰 줘서 잘 구웠지.

<IN THE SOOP BTS ver. Season 2> EP. 4

Jin and Jung Kook think about the time they played *jokgu* together in the rain   빗속에서 함께 족구했던 시간을 떠올리는 진과 정국

On rainy days when both your mind and body feel drowsy, try this simple *kimchijeon* recipe and pair it with a drink. It does not have to be *makgeolli.* The crispy *kimchijeon* will make even the humid air crisp, and one refreshing sip of the drink will wash away the greasiness of *jeon*.

비가 내려 몸과 마음이 나른한 날에는 간단하게 만든 김치전에 술을 한 잔 곁들여 보자. 막걸리가 아니어도 괜찮다. 바삭한 김치전이 눅눅한 공기마저 바삭하게 만들어 주고, 가벼운 술 한 모금이 전의 기름기를 깔끔하게 씻어내는 느낌을 줄 것이다.

RECIPE →

page 206

BTS *Choice*

(Playing the liar game*)

Mission word: *kimchijeon*

(*라이어 게임 중) 제시어: 김치전

<Run BTS!> EP. 128

Jimin

It's a simple dish. 단순한 음식이에요.

RM

It tastes best when it's warm.
이 음식은 따뜻할 때 먹어야 해요.

Jung Kook

This dish is very red. 이 음식은 굉장히 빨개요.

Jin

(Liar) Most people like this dish.
(라이어) 대부분 사람이 호불호 없이 좋아해요.

V

You have to use chopsticks (to eat this dish).
이 음식은 (먹을 때) 젓가락을 사용해야 합니다.

j-hope

I made this dish and ate it recently.
이 음식 제가 최근에 해 먹어 봤습니다.

*SUGA did not attend due to personal reasons 슈가는 개인 사정으로 불참

---

* The "liar game" is a Korean word game. When the game starts, everyone receives the same mission word (e.g. *kimchijeon*) except for the liar, who gets the word "liar" instead of the mission word. Without knowing who the liar is, the players try to find the liar as they talk about the mission word.

라이어 게임은 한국의 말놀이 중 하나다. 게임이 시작되면 라이어를 제외한 참가자들은 모두 같은 제시어(예. 김치전)를 받고, 라이어는 제시어 대신 '라이어'라는 단어를 받는다. 참가자들은 라이어가 누구인지 모르는 채로 제시어에 관해 이야기를 나누며 라이어를 찾아낸다.

A Pinch *of* Korean

짠! 건배!

Saying "Cheers" in Korean

When making a toast in Korea, you say "geon-bae" (건배), but you can also say "jjan" (짠) with close friends. 짠 is a word that describes the sound of glasses clinking, much like the word "clink" in English. Propose a toast with your close friends and see if you can hear the sound "jjan" (짠) as you clink your glasses!

"BTS (방탄소년단) '찐 방탄회식' #2022BTSFESTA"

- PERFECT PAIRINGS WITH DRINKS -

Gyeran Mari

Rolled Omelet

계란말이

#Mild #Rolled #Eggs

RECIPE →

page 208

Following Jin's instructions, V and Jung Kook finished making cheese *gyeran mari* (rolled omelet) in 10 minutes on the "Avatar Chef King" episode of <Run BTS!>. Jin gave speedy directions with added sound effects like "chop-chop," "sizzle," and "fizzle" during his chance to help V and Jung Kook cook. It must have been quite helpful! V and Jung Kook transferred the finished *gyeran mari* to a plate and added the finishing touches with a drizzle of ketchup.

V, dubbed "Kim Deco" for his aesthetic sense, even added some plastic fruit and a flowerpot as decorations around the plate. And he didn't forget to pay lip service to j-hope, who was the game's emcee and judge of the day.

<달려라 방탄> '아바타 셰프왕' 편에서 뷔와 정국은 진의 지시에 따라 10분 만에 치즈 계란말이를 만들어냈다. 찬스를 이용해 잠시 둘을 도와주러 온 진이 착착착, 차아악, 치이익 등 효과음을 더해 속성으로 알려 주었던 설명이 도움이 되었던 모양이다. 뷔와 정국은 빠르게 완성한 계란말이를 접시에 올린 다음 케첩까지 뿌려 마무리했다.

게다가 꾸미는 감각이 있어 '김데코'라는 별명이 붙은 뷔는 과일 모형과 화분을 가져와 접시 주변까지 열심히 꾸몄다. 그리고 이날의 MC이자 심사위원이었던 제이홉에게 달콤한 멘트를 한 마디 던지는 것도 잊지 않았다.

<Run BTS!> EP. 102

V and Jung Kook make *gyeran mari* 계란말이를 만드는 뷔와 정국

V decorates the dish with a flowerpot and fruit basket 뷔가 화분과 과일 바구니를 사용하여 요리를 꾸민 모습

Jin

For gyeran mari,
scallions go
"chop-chop,"
and the egg mixture goes
"sizzle!"

계란말이는 어떻게 하냐면,
쪽파를 얇게 착착착 썰어 넣고,
프라이팬 위에 차아악~

<Run BTS!> EP. 102

V

You look so beautiful,
even when you eat.

먹는 모습마저 예쁘시네요.

Gyeran mari is a dish made by rolling up a thin layer of cooked eggs. A typical Korean-style *gyeran mari* is made by mixing finely chopped vegetables, such as carrots, scallions, and onions, with eggs and frying the mixture. For a more special *gyeran mari*, you can add sliced cheese, just like V and Jung Kook, or even add ham or crab sticks. Lastly, you can drizzle ketchup on the finished *gyeran mari* to add a sweet and sour flavor.

계란말이는 이름 그대로 얇게 부친 계란을 말아서 만드는 요리다. 일반적으로 한국식 계란말이는 계란물에 당근과 쪽파, 양파 등 잘게 다진 채소를 섞어 알록달록한 색이 나도록 굽는다. 조금 더 특별한 계란말이를 만들고 싶다면 뷔와 정국처럼 슬라이스 치즈를 넣거나, 햄이나 맛살을 추가해도 맛있다. 마지막으로 완성된 계란말이에 케첩을 뿌리면 새콤달콤한 맛을 더할 수 있다.

<Run BTS!> EP. 102

Jimin, SUGA, and RM watch V and Jung Kook from the situation room
상황실에서 뷔와 정국을 지켜보는 지민, 슈가, RM

Jin and j-hope watch V and Jung Kook from the situation room 상황실에서 뷔와 정국을 지켜보는 진과 제이홉

If you are looking for a light side dish or a quick late-night snack, why don't you try making some delicious, easy-to-make *gyeran mari*? It is not only a popular side dish for rice, but it is also frequently enjoyed with alcohol. When you don't feel like pairing a greasy dish with your drink, *gyeran mari* will be a great option for you.

RECIPE →
page 208

담백한 곁들임 메뉴를 고민 중이거나 늦은 시간 간단하게 먹을 요리를 찾고 있다면 만들기 쉽고 맛도 좋은 계란말이에 도전해 보면 어떨까? 계란말이는 밥반찬으로도 사랑받지만, 술과 함께 즐기는 경우도 많다. 기름진 안주가 부담스러울 때, 계란말이도 좋은 선택지가 되어 줄 것이다.

<Run BTS!> EP. 102

V and Jung Kook make preparations before cooking 요리 시작 전, 준비 중인 뷔와 정국

A Pinch *of* Korean

말이
구이
볶이

Ma-ri, Gu-i, Bo-kki

The names of Korean dishes sometimes include the main cooking method used to make them. An example of this is *gyeran mari* (계란말이; rolled omelet), which includes the word ma-ri (말이), which describes a rolled dish. Let's learn more about the names of Korean dishes with similar structures.

Infinitives		Conjugations		Dishes
mal-da 말다 to roll	→	*ma-ri* 말이 rolled dish	·········	*gye-ran-ma-ri* 계란말이 rolled omelet
gup-da 굽다 to grill	→	*gu-i* 구이 grilled dish	·········	*gal-bi-gu-i* 갈비구이 grilled *galbi* (ribs)
bok-da 볶다 to stir-fry	→	*bo-kki* 볶이 stir-fried dish	·········	*tteok-bo-kki* 떡볶이 stir-fried rice cakes

7

Sweet Desserts

달달한 디저트

Dalgona 달고나

Hwachae 화채

- SWEET DESSERTS -

Dalgona

Sugar Candy

달고나

RECIPE ⟶

page 212

Dalgona (sugar candy) is a street food that BTS frequently enjoyed as a snack during their schooldays. It is also called *jjokja, ppopgi,* and *ttigi* depending on the region, but the ingredients and recipes for making them are almost identical. The only ingredients you need for *dalgona* are sugar and baking soda! Making *dalgona* may look easy because of its simple ingredients, but don't be fooled. It actually requires a bit of skill and precise timing. On <Run BTS!>, some of the members who were not satisfied with the results of their first attempt were eager to try again to make the perfect *dalgona*.

달고나는 방탄소년단이 학창 시절에 간식으로 자주 먹었던 길거리 음식이다. 달고나는 달고나라는 이름 외에도 지역에 따라 쪽자, 뽑기, 띠기 등 다양한 이름으로 불리는데, 만드는 재료와 방식은 거의 동일하다. 달고나를 만들기 위해 필요한 재료는 설탕과 베이킹 소다뿐! 재료가 단순해서 간단해 보이지만, 정확한 타이밍과 약간의 기술이 요구되는 요리이므로 방심해서는 안 된다. <달려라 방탄>에서 첫 도전의 결과물이 만족스럽지 않았던 일부 멤버들은 달고나 만들기에 다시 도전하는 열정을 보이기도 했다.

<Run BTS!> EP. 95

BTS makes *dalgona* together　사이좋게 달고나를 만드는 방탄소년단

j-hope

Don't you smell something burning?

어디서 타는 냄새 나지 않아요?

Jung Kook

Is it the smell of my burning love?

내 마음이 타는 냄새?

<Run BTS!> EP. 95

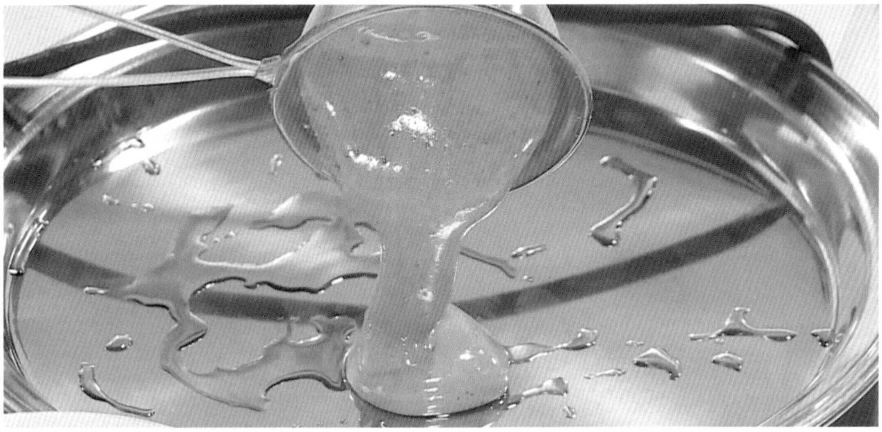

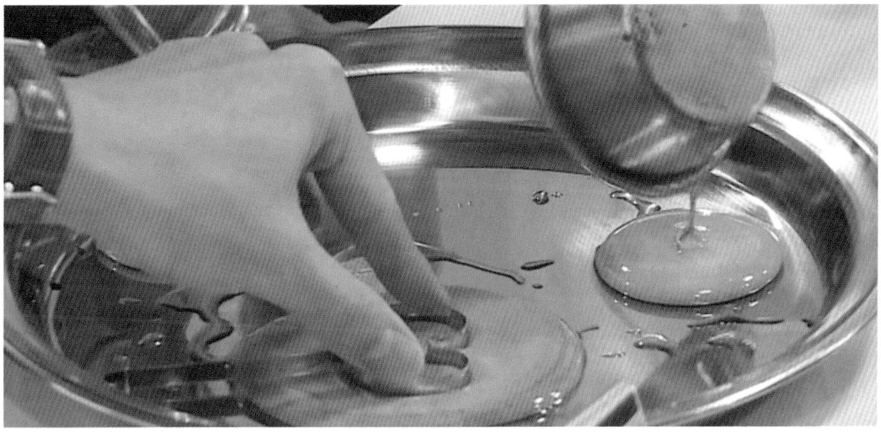

The *dalgona* is coming together quickly 빠르게 완성되어 가는 달고나

Dalgona is delicious once it has been cooled, but it is not truly complete until it has been flattened on a tray and stamped with a shape. Common shapes for *dalgona* include hearts, stars, and umbrellas. In the old days, after buying these stamped *dalgona*, if you succeeded in picking out the inner shape without cracking it, you could show it to the store and get another *dalgona* for free. Though *dalgona* looks hard on the outside, it breaks easily under the slightest pressure. Therefore, to pick out the shapes with precision, you can use tools like needles or toothpicks. On <Run BTS!>, BTS also played this game where they had to pick out the shapes using needles. Their intense concentration while picking out the *dalgona* shapes showed their sincere dedication toward the game.

달고나는 잘 식혀서 그냥 먹어도 맛있지만, 쟁반에 올려 납작하게 누른 다음 모양까지 찍어 줘야 완성이라고 할 수 있다. 기본 모양틀은 하트, 별, 우산 등 여러 가지인데, 옛날에는 모양이 찍힌 달고나를 산 뒤, 모양에 맞게 달고나를 잘 떼어내는 데에 성공할 경우, 달고나를 샀던 가게에 가져가 보여 주면 새 달고나를 덤으로 받을 수 있었다. 달고나는 겉보기에 단단해 보이지만 조금만 힘을 주어도 잘 부서진다. 때문에 보다 정교하게 모양을 떼어낼 때는 바늘이나 이쑤시개 같은 도구를 이용하기도 한다. <달려라 방탄>에서 방탄소년단 역시 바늘을 사용하여 달고나를 떼어내는 게임을 했는데, 각자 집중력을 발휘하는 모습에서 게임에 대한 그들의 진심이 잘 드러났다.

RECIPE →
page 212

<Run BTS!> EP. 95

V, Jung Kook, and j-hope focus on removing the edges around the *dalgona* shapes
달고나를 모양대로 떼어내기 위해 집중하는 뷔, 정국과 제이홉

Jimin and j-hope try to make *dalgona* again　달고나 만들기에 재도전하는 지민과 제이홉

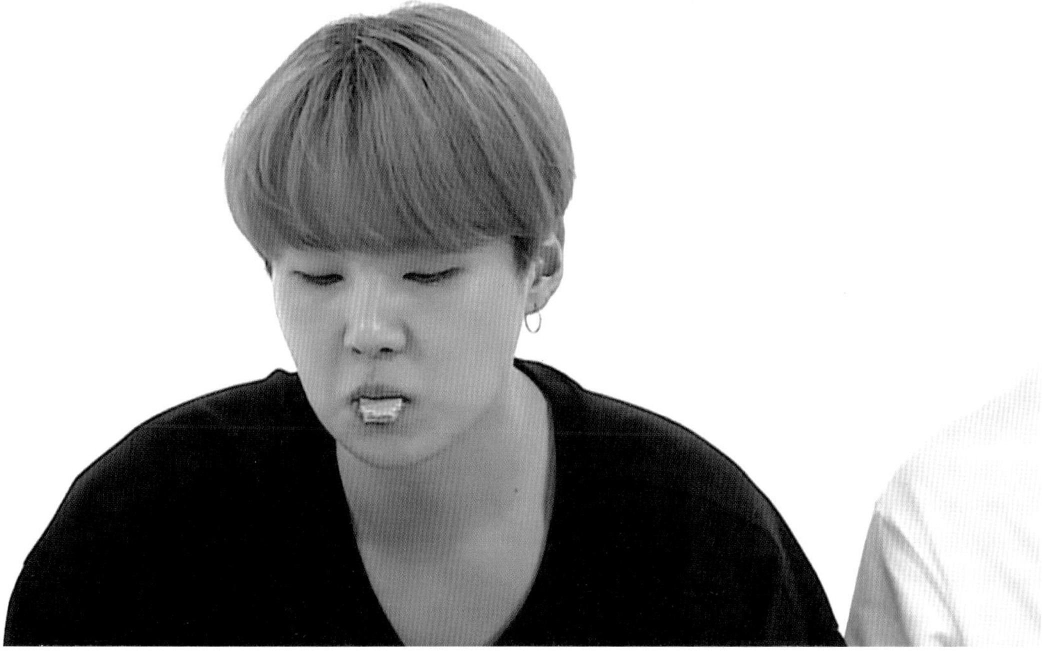

SUGA enjoys the *dalgona* he made　직접 만든 달고나를 맛있게 먹는 슈가

SUGA

I ate a lot of dalgona as a kid.

나 어릴 때 달고나 엄청 먹었어.

<Run BTS!> EP. 95

A Pinch *of* Korean

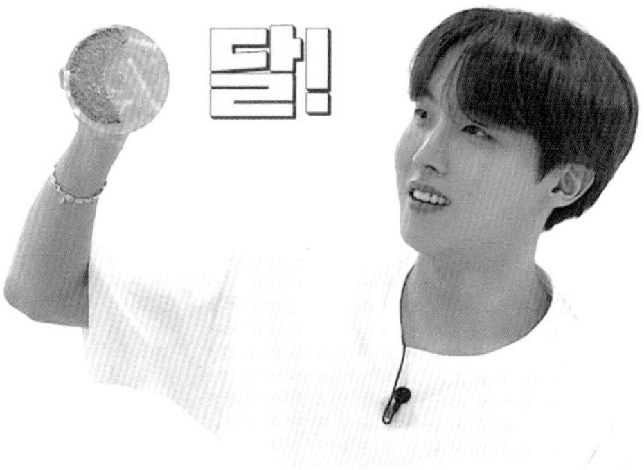

Korean Expressions for Taste

It is said that "dal" (달) of *dalgona* comes from the word "dal-da" (달다; to be sweet), an adjective that describes taste. The expressions below are Korean adjectives that describe tastes.

Adjectives Describing Tastes

dal-da
달다
to be sweet

shi-da
시다
to be sour

maep-da
맵다
to be spicy

jja-da
짜다
to be salty

sseu-da
쓰다
to be bitter

*Spiciness is closer to a sense of pain than a kind of taste.
But in Korea, it is often expressed as a taste.

- SWEET DESSERTS -

Hwachae

Punch

화채

#Refreshing #Mixed #Dessert

RECIPE →

page 214

One hot summer day, after eating their first meal on <IN THE SOOP BTS ver. Season 2>, BTS decided to have some refreshing *hwachae* (punch) for dessert. *Hwachae* is a sweet, chilled Korean summer drink. Typically, a variety of fruit cut into bite-sized pieces is added to a drink made of a carbonated drink. A wide selection of fruit was available, including a watermelon, some bananas, and some apples, but the main ingredient that SUGA chose was watermelon. SUGA first cut the large watermelon in half and started to scoop out half of it with a spoon. To save the other half of the watermelon for later, Jung Kook asked RM to cover it with plastic wrap. Instead, RM performed a quick rap as he looked at the watermelon. This was a witty pun derived from the same pronunciations of wrap and rap.

무더운 여름, <인더숲 BTS편 시즌 2>에서의 첫 식사를 마친 방탄소년단은 후식으로 시원한 화채를 먹기로 했다. 화채는 달콤하고 시원한 한국의 여름 음료로, 한입 크기로 자른 다양한 과일을 탄산음료에 넣어 만든다. 숙소에 있는 수박, 바나나, 사과 등 다양한 과일 중, 슈가가 선택한 주재료는 수박이었다. 슈가는 우선 커다란 수박을 반으로 자르고, 숟가락으로 수박 속을 파내기 시작했다. 남은 수박 반 통은 비닐 랩을 씌워 보관하기 위해 정국이 RM에게 수박에 비닐 랩을 씌워 달라고 말했는데, 그 말을 들은 RM은 수박을 바라보며 짧은 랩을 해 주었다. 두 '랩'의 발음이 같다는 점을 이용한 재치 있는 농담이었다.

<IN THE SOOP BTS ver. Season 2> EP. 2

j-hope, Jung Kook, and SUGA make *hwachae* 화채를 만드는 제이홉, 정국과 슈가

Jung Kook

RM, could you cover this with plastic wrap?

남준이 형, 여기 랩 좀 씌워 주세요.

RM

(starts to rap) Yo! Check it!

(랩을 하기 시작한다) Yo! Check it!

<IN THE SOOP BTS ver. Season 2> EP. 2

Picking off where SUGA left off, j-hope and Jung Kook finished scooping out the watermelon, and SUGA went on to fill a large bowl by mixing watermelon, canned peaches, a carbonated drink, milk, and ice together. When the *hwachae* was ready, everyone in the room —including j-hope, who had said he had wanted to eat *hwachae* ever since his <IN THE SOOP BTS ver. Season 2> interview filmed before the show—emptied a bowl of the cool dessert. After enjoying this tasty dessert together, everyone leisurely pondered what they wanted to do that afternoon. It was truly the start of BTS' summer vacation.

슈가에 이어 제이홉과 정국이 수박 반 통의 속을 모두 파내자, 슈가는 수박과 복숭아 통조림, 탄산음료, 우유와 얼음을 잘 섞어 커다란 볼을 가득 채웠다. 화채가 완성되자 <인더숲 BTS편 시즌 2> 사전 인터뷰 때부터 화채를 먹고 싶다고 말했던 제이홉을 포함하여 자리에 있던 모두가 시원하게 한 그릇씩을 비워 냈다. 함께 맛있는 후식을 먹고, 각자 하고 싶은 것을 고민하는 한가로운 오후. 그야말로 방탄소년단의 여름휴가가 시작되는 순간이었다.

RECIPE ⟶
page 214

RM, Jin, SUGA, j-hope, and Jung Kook enjoy the finished *hwachae*

완성된 화채를 맛있게 먹는 RM, 진, 슈가, 제이홉과 정국

<IN THE SOOP BTS ver. Season 2> EP. 2

Jimin enjoys playing basketball alone while the others eat *hwachae*
다른 멤버들이 화채를 먹는 동안 혼자 농구를 즐기는 지민

V takes an online trumpet lesson while the others eat *hwachae*
다른 멤버들이 화채를 먹는 동안 온라인 트럼펫 레슨을 받는 뷔

BTS relaxes in their own way after eating *hwachae*　화채를 먹은 후 각자의 방식으로 휴식을 취하는 방탄소년단

Weverse (2020.12.28)

<Run BTS!> EP. 131

BTS *Choice*

Q.

Hard Peaches vs. Soft Peaches

딱딱한 복숭아 vs 물렁물렁한 복숭아?

Weverse (2020.12.28) & <Run BTS!> EP. 131

SUGA

I like both. 저는 둘 다 좋아해요.

V

Soft peaches, of course!
물렁한 복숭아지!

Jung Kook

Definitely soft peaches! 물. 복.이죠!

Q.

Fruit vs. Bread

과일 vs 빵?

<Run BTS!> EP. 93

Jin, V

Fruit! 과일!

RM

Bread! 빵!

A Pinch *of* Korean

자르다
·
푸다
·
섞다

Ja-reu-da, Pu-da, Seok-da

Let's think about the process SUGA, j-hope, and Jung Kook went through to make *hwachae* and learn about Korean verbs for cooking.

Verbs Used When Making *Hwachae*

ja-reu-da	*pu-da*	*seok-da*
자르다	푸다	섞다
to cut	to scoop	to mix

Recipe

1
Easy Homemade Meals
간편한 집밥

2
Box Lunches Just for You
취향 따라 만드는 도시락

3
Slurp! Noodle Dishes
후루룩 면 요리

4
Spicy Meat Dishes
매콤한 고기 요리

5
Dishes for Special Feasts
특별한 만찬

6
Perfect Pairings with Drinks
술과 어울리는 안주

7
Sweet Desserts
달달한 디저트

Basic Preparations
기본 조리법

1

Easy Homemade Meals

간편한 집밥

Ganjang Gyeran Bap 간장계란밥

Kimchi Jjigae 김치찌개

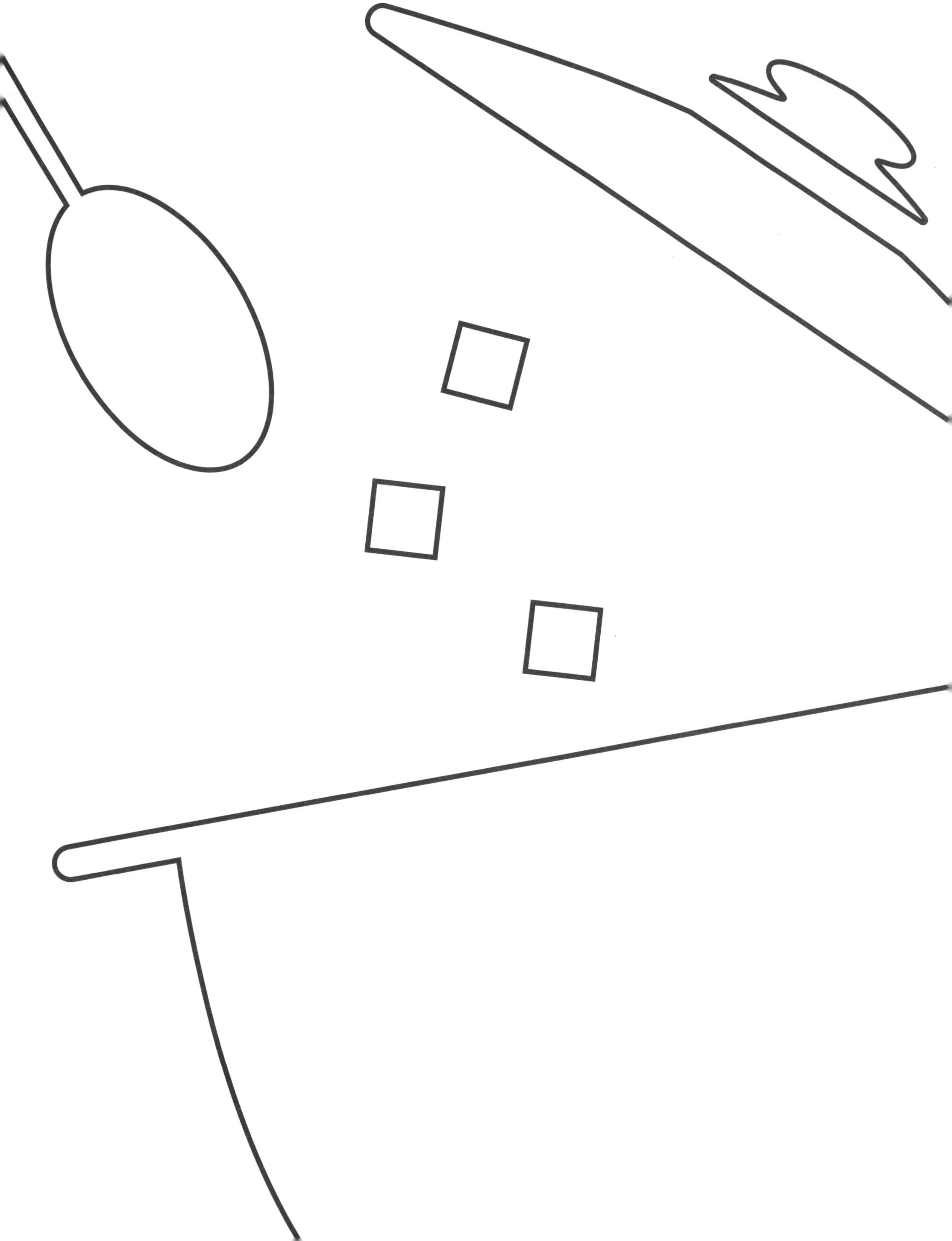

Serves **1**
1인분

Difficulty **Low**
난이도 **쉬움**

- EASY HOMEMADE MEALS -

Ganjang Gyeran Bap

Egg Rice with Soy Sauce

간장계란밥

STORY ⟶ page **20**

INGREDIENTS

- ☐ 2 **eggs**
- ☐ 2 oz / 60 g **beef sirloin**
- ☐ 2 (0.3 oz / 10 g) **scallions**
- ☐ 2 tablespoons **soy sauce**
- ☐ 2 tablespoons **toasted sesame oil**
- ☐ ½ teaspoon **toasted sesame seeds**
- ☐ ⅓ teaspoon **salt**
- ☐ 1 to 2 tablespoons **cooking oil**

RICE (7 oz / 200 g)

- ☐ ½ cup (4 oz / 110 g) **uncooked rice**
- ☐ ¾ cup (5 oz / 150 ml) **water**

- ☐ **계란** 2개
- ☐ **소고기 등심** 60g
- ☐ **쪽파** 2대 (10g)
- ☐ **간장** 2큰술
- ☐ **참기름** 2큰술
- ☐ **참깨** ½작은술
- ☐ **소금** ⅓작은술
- ☐ **식용유** 1-2큰술

밥 (200g)

- ☐ **쌀** ½컵 (110g)
- ☐ **물** ¾컵 (150ml)

Vegan 비건

- ☐ 3.5 oz / 100 g **vegan eggs**
- ☐ 2 oz / 60 g **tofu**

- ☐ **비건 계란** 100g
- ☐ **두부** 60g

Gluten-free 글루텐 프리

- ☐ 2 tablespoons **gluten-free soy sauce**

- ☐ **글루텐 프리 간장** 2큰술

Tips

- You can also use other fatty beef cuts instead of sirloin. But this dish is delicious even if you make it with just fried eggs without the beef.
- Adding scallions and toasted sesame seeds can enrich the flavor and texture, but you can leave them out if they are unavailable.

- 소고기는 등심 대신 지방이 많은 다른 구이용 부위를 사용해도 된다. 하지만 소고기를 제외하고 계란프라이만 넣어 만들어도 맛있다.
- 쪽파와 참깨를 넣으면 식감과 풍미가 더 좋아지지만, 구하기 어렵다면 제외해도 된다.

HOW TO COOK

1. Prepare the rice. → **page 218**
2. Finely chop the scallions.
3. Pat the beef dry with a paper towel and chop it into bite-sized pieces.
4. Add the cooking oil to a skillet and heat it. Crack the eggs, and fry them over easy or over hard depending on your preference. Then transfer them to a plate.

   Tip How to make over-easy or over-hard eggs

   Add the cooking oil to a skillet and heat it over high heat. Crack an egg into the skillet and lower the heat to medium. Once the white turns opaque, flip the egg with a spatula. Cook it for 1 to 2 minutes over low heat to make it over easy. Cook it for 3 minutes over medium heat to make it over hard.
5. Wipe the grease off the skillet and grill the chopped beef on the dry skillet. Sprinkle some salt on the beef as you grill it.

   Vegan Use tofu instead of beef. Wipe the moisture off the tofu with a paper towel. Slice it into 0.4-in/1-cm thick cubes. Add the cooking oil to a skillet and heat it. Cook the tofu over medium heat until it is golden brown. Sprinkle some salt on the tofu as you grill it.
6. Put the rice into a bowl and top it with the fried eggs, grilled beef, finely chopped scallions, soy sauce, toasted sesame oil, and toasted sesame seeds.

1. 밥을 준비한다. → **218 쪽**
2. 쪽파를 잘게 다진다.
3. 소고기는 키친타월을 사용하여 핏물을 닦아 낸 뒤, 먹기 좋은 크기로 썬다.
4. 팬에 식용유를 두르고 불에 달군 뒤, 계란을 깨뜨려 넣고 취향에 따라 반숙 혹은 완숙으로 프라이를 만든다. 완성된 프라이는 접시로 옮겨 둔다.

   Tip **계란프라이 반숙, 완숙 만들기**

   팬에 식용유를 두르고 강불로 달군다. 계란을 깨뜨려 넣고 중불로 줄인다. 흰자가 불투명해졌을 때 뒤집개로 뒤집어 약불에 1-2분간 두면 반숙이 되고, 뒤집은 계란을 중불에 3분간 두면 완숙이 된다.
5. 팬의 기름을 닦아내고, 마른 팬에 썰어 둔 소고기를 굽는다. 굽는 사이사이 소금을 뿌린다.

   Vegan 소고기 대신 두부를 사용한다. 두부의 물기를 키친타월로 닦아 내고, 1cm 두께로 네모나게 썬다. 팬에 식용유를 두르고 달군 뒤, 두부를 넣고 중불에서 노릇하게 굽는다. 굽는 사이사이 소금을 뿌린다.
6. 밥을 그릇에 담고, 계란프라이, 구운 소고기, 다진 쪽파, 간장, 참기름, 참깨를 얹는다.

Special Tip <BTS BON VOYAGE Season 4> EP. 8

By using the Korean-style soy sauce and toasted sesame oil, you will be able to recreate the same nostalgic taste that we used to enjoy!
한국식 간장과 참기름을 사용하면 저희가 먹었던 추억의 맛을 그대로 구현할 수 있을 거예요!

Serves **2**
2인분

Difficulty **Mid**
난이도 **보통**

- EASY HOMEMADE MEALS -

Kimchi Jjigae

Kimchi Stew

김치찌개

STORY → page **28**

INGREDIENTS

- ☐ 1⅔ cups (14 oz / 400 g) kimchi
- ☐ 2 strips (10 oz / 300 g) pork belly (*samgyeopsal*)
- ☐ ⅓ (2.5 oz / 70 g) onion
- ☐ ¼ (4 in / 10 cm, 1 oz / 30 g) large green onion
- ☐ 4 cups (28 oz / 800 ml) water
- ☐ ¾ cup (5 oz / 150 ml) kimchi juice
- ☐ 1 tablespoon salt
- ☐ 1 teaspoon minced garlic
- ☐ ½ teaspoon sugar
- ☐ 1 tablespoon cooking oil

- ☐ **김치** 1⅔컵 (400g)
- ☐ **돼지고기 삼겹살** 2줄 (300g)
- ☐ **양파** ⅓개 (70g)
- ☐ **대파** ¼대 (10cm, 30g)
- ☐ **물** 4컵 (800ml)
- ☐ **김칫국물** ¾컵 (150ml)
- ☐ **소금** 1큰술
- ☐ **다진 마늘** 1작은술
- ☐ **설탕** ½작은술
- ☐ **식용유** 1큰술

Tips

- *Kimchi jjigae* tastes best when made with well-fermented kimchi. Well-fermented kimchi has a sour taste. If your kimchi is not sour enough, add 1 teaspoon of vinegar to the stew before serving it.
- Adding tofu that is sliced into bite-sized pieces is also a good idea.

- 김치찌개는 잘 익은 김치로 끓여야 맛있다. 잘 익은 김치는 새콤한 맛이 나는데, 이 맛이 부족하다면 먹기 직전에 식초를 1작은술 추가한다.
- 두부를 먹기 좋은 크기로 썰어 추가해도 잘 어울린다.

Vegan 비건

- ☐ 1⅔ cups (14 oz / 400 g) vegan kimchi
- ☐ 3.5 oz / 100 g tofu
- ☐ 2 (1.5 oz / 40 g) oyster mushrooms
- ☐ ¾ cup (5 oz / 150 ml) vegan kimchi juice

- ☐ **비건 김치** 1⅔컵 (400g)
- ☐ **두부** 100g
- ☐ **느타리버섯** 2개 (40g)
- ☐ **비건 김칫국물** ¾컵 (150ml)

Gluten-free 글루텐 프리

- ☐ 1⅔ cups (14 oz / 400 g) gluten-free kimchi
- ☐ ¾ cup (5 oz / 150 ml) gluten-free kimchi juice

- ☐ **글루텐 프리 김치** 1⅔컵 (400g)
- ☐ **글루텐 프리 김칫국물** ¾컵 (150ml)

HOW TO COOK

1. Julienne the onion into 0.2-in/0.5-cm thick strips, and slice the large green onion diagonally into 0.1-in/0.2-cm thick pieces. Then cut the kimchi and pork belly into bite-sized pieces.

   Vegan Use vegan kimchi and leave out the pork belly. Drain the tofu and cut it into bite-sized pieces. Cut off the bottom of the oyster mushrooms and tear them into strips by hand.

2. Add the cooking oil to a pot and heat it. Add the kimchi and pork belly and stir-fry over medium heat for 5 minutes. Add water and bring it to a boil.

   Tip Use the same amount of vegetable, anchovy, or beef broth instead of water to enrich the flavor. → **pages 219-221**

3. When the water starts boiling, add the onion, minced garlic, and kimchi juice. Let it simmer for 30 minutes over medium heat (make sure not to add the large green onion yet).

4. Season with sugar and salt to taste and let it simmer for 10 more minutes over low heat.

   Tip Add sugar and salt in small amounts at a time, as the flavor may vary depending on the kimchi.

   Vegan Season with sugar and salt and add the prepared tofu and oyster mushrooms. Let it simmer for 10 more minutes over low heat.

5. Lastly, add the sliced large green onion and turn off the heat.

1. 양파는 0.5cm 두께로 채 썰고, 대파는 0.2cm 두께로 어슷하게 썬다. 김치와 삼겹살은 먹기 좋은 크기로 썬다.

   Vegan 비건 김치를 사용하고, 삼겹살은 제외한다. 두부는 물기를 제거하고 먹기 좋은 크기로 자른다. 느타리버섯은 밑동을 잘라내고 결을 따라 손으로 찢는다.

2. 냄비에 식용유를 두르고 불에 달군 뒤, 김치와 삼겹살을 넣어 중불에서 5분간 볶는다. 냄비에 물을 넣고 끓인다.

   Tip 물 대신 같은 분량의 채소 국물이나 멸칫국물, 고깃국물을 사용하면 맛이 더 풍부해진다. → **219-221 쪽**

3. 물이 끓으면 냄비에 양파, 다진 마늘, 김칫국물을 넣고 중불에서 30분간 끓인다. (대파는 아직 넣지 않는다.)

4. 설탕과 소금을 넣어 간을 맞추고 약불에서 10분간 더 끓인다.

   Tip 김치에 따라 간이 다를 수 있으므로 설탕과 소금은 조금씩 넣어 가며 간을 맞춘다.

   Vegan 설탕과 소금을 넣어 간을 맞춘 다음, 손질해 둔 두부와 느타리버섯을 넣고 약불에서 10분간 더 끓인다.

5. 마지막으로 썰어 둔 대파를 넣고 불을 끈다.

Special Tip <BTS BON VOYAGE Season 4> EP. 4

Add about ½ cup (3.5 oz / 100 ml) more water to the leftover broth to enjoy with *ramyeon* noodles (instant noodles) and sausage.
남은 국물에 물을 ½컵 (100ml) 정도 더 넣어 라면 사리와 소시지를 추가해도 잘 어울려요.

2

Box Lunches Just for You

취향 따라 만드는 도시락

Gimbap 김밥

Toast 토스트

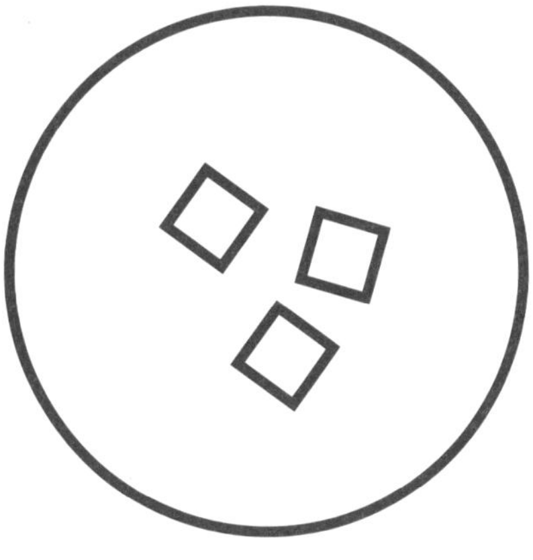

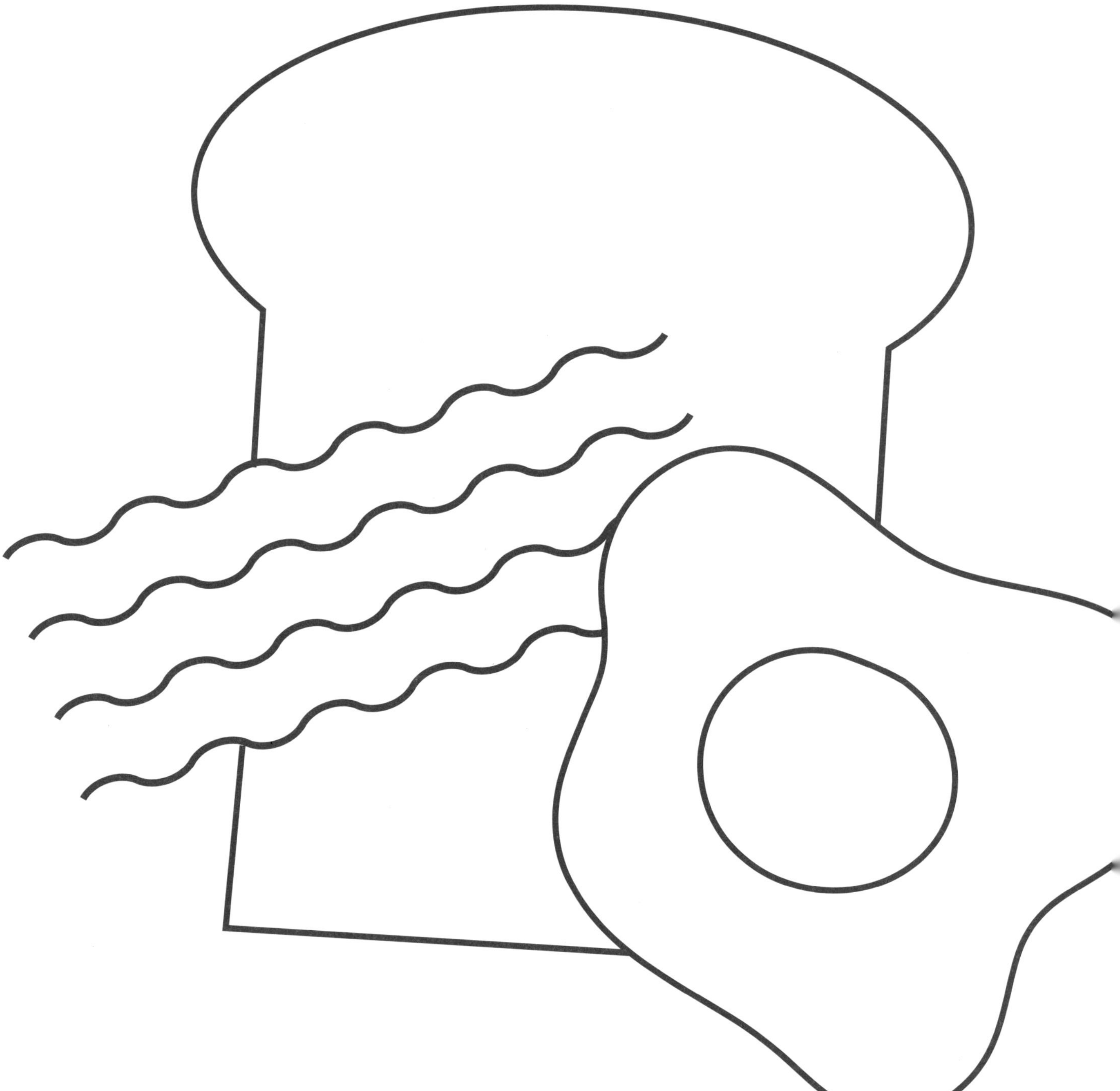

Serves **2** (2 rolls)
2인분(2줄)

Difficulty **Mid**
난이도 **보통**

- BOX LUNCHES JUST FOR YOU -

Gimbap

Seaweed Rice Rolls

김밥

STORY ⟶ page **38**

INGREDIENTS

- ☐ ⅓ (3.3 oz / 90 g) cucumber
- ☐ ¼ (3.3 oz / 90 g) carrot
- ☐ 3.3 oz / 90 g burdock root
- ☐ 2 eggs
- ☐ 2 strips (1.3 oz / 36 g) pickled radish
- ☐ 2 strips (1 oz / 32 g) ham
- ☐ 2 sheets (7.5 in × 8.3 in / 19 cm × 21 cm) *gim* (dried seaweed)
- ☐ 1 teaspoon toasted sesame oil
- ☐ ⅓ teaspoon salt
- ☐ 7 tablespoons cooking oil

- ☐ **오이** ⅓개 (90g)
- ☐ **당근** ¼개 (90g)
- ☐ **우엉** 90g
- ☐ **계란** 2개
- ☐ **단무지** 2줄 (36g)
- ☐ **햄** 2줄 (32g)
- ☐ **김** 2장 (19cm × 21cm)
- ☐ **참기름** 1작은술
- ☐ **소금** ⅓작은술
- ☐ **식용유** 7큰술

RICE (14 oz / 400 g)

- ☐ 1⅕ cups (7.7 oz / 220 g) uncooked rice
- ☐ 1½ cups (10 oz / 300 ml) water

밥 (14 oz / 400 g)

- ☐ **쌀** 1⅕컵 (220g)
- ☐ **물** 1½컵 (300ml)

SEASONING FOR RICE

- ☐ 2 teaspoons toasted sesame oil
- ☐ 1 teaspoon salt

밥 양념

- ☐ **참기름** 2작은술
- ☐ **소금** 1작은술

SEASONING FOR BURDOCK ROOT

- ☐ 1½ tablespoons soy sauce
- ☐ 1 tablespoon *mullyeot* (corn syrup)
- ☐ ½ tablespoon sugar

우엉 양념

- ☐ **간장** 1½큰술
- ☐ **물엿** 1큰술
- ☐ **설탕** ½큰술

PORK BELLY GIMBAP

- ☐ 1 strip (5 oz / 150 g) pork belly (*samgyeopsal*)
- ☐ 1 (0.3 oz / 10 g) chili pepper
- ☐ 1 tablespoon *ssamjang* (red chili and soybean paste)

삼겹살 김밥

- ☐ **돼지고기 삼겹살** 1줄 (150g)
- ☐ **고추** 1개 (10g)
- ☐ **쌈장** 1큰술

Vegan 비건

- ☐ 2 strips (1 oz / 32 g) vegan ham

- ☐ **비건 햄** 2줄 (32g)

Gluten-free 글루텐 프리

- ☐ 2 strips (1 oz / 32 g) gluten-free ham
- ☐ 1½ tablespoon gluten-free soy sauce (for burdock root seasoning)
- ☐ 1 tablespoon gluten-free *ssamjang* (for pork belly *gimbap*)

- ☐ **글루텐 프리 햄** 2줄 (32g)
- ☐ **글루텐 프리 간장** 1½큰술 (우엉 양념)
- ☐ **글루텐 프리 쌈장** 1큰술 (삼겹살 김밥)

Tips

- If burdock root is unavailable, you can replace it with ½ (3.5 oz / 100 g) julienned onion. In this case, use only 1⅓ tablespoons of soy sauce and 2 teaspoons of sugar for the seasoning.
- It is convenient to use pre-cut pickled radish and ham sliced lengthwise for *gimbap*. If they are unavailable, use regular pickled radish and ham and cut them into 0.4-in/1-cm thick strips.
- If *ssamjang* is unavailable, you can easily make it by mixing 1 teaspoon of *gochujang* (Korean red chili paste), 1 teaspoon of *doenjang* (soybean paste), 1 tablespoon of water, and ½ teaspoon of sugar.
- If you add the ingredients below in addition to the basic *gimbap* ingredients (cucumber, carrot, burdock root, egg, pickled radish, ham), you can make many kinds of *gimbap*.
  - Tuna *gimbap*: 3.5 oz / 100 g oil-removed canned tuna, 4 perilla leaves (*kkaennip*), 2 tablespoons mayonnaise
  - Kimchi *gimbap*: ½ cup (4.3 oz / 120 g) of drained and finely chopped kimchi (It tastes better when you mix ½ tablespoon of toasted sesame oil with the kimchi.)
  - Cheese *gimbap*: 2 slices of cheddar cheese
  - Keto (low-carbohydrate) *gimbap*: 1 sheet of shredded egg instead of rice

- 우엉을 구하기 어렵다면 우엉 대신 양파 ½개(100g)를 채 썰어 사용해도 된다. 이때, 양념 재료로는 간장 1⅓큰술과 설탕 2작은술만 사용한다.
- 단무지와 햄은 세로로 긴 김밥용 재료를 사용하면 편리하다. 김밥용 단무지나 햄이 없다면 일반 단무지나 햄을 1cm 두께로 길게 썰어 준비한다.
- 쌈장을 구하기 어렵다면 고추장 1작은술, 된장 1작은술, 물 1큰술, 설탕 ½작은술을 고루 섞어 쉽게 만들 수 있다.
- 기본 김밥 재료(오이, 당근, 우엉, 계란, 단무지, 햄)에 아래 재료를 추가하면 다양한 김밥을 만들 수 있다.
  - **참치김밥**: 기름을 제거한 캔 참치 100g, 깻잎 4장, 마요네즈 2큰술
  - **김치김밥**: 물기를 제거하고 잘게 다진 김치 ½컵 (120g) (참기름 ½큰술을 김치에 넣고 잘 버무려 사용하면 더 맛있다.)
  - **치즈김밥**: 슬라이스 체더 치즈 2장
  - **키토(저탄수화물) 김밥**: 밥 대신 잘게 채 썬 계란지단 1장

HOW TO COOK

1. Prepare the rice. → **page 218**
2. Slice the cucumber lengthwise into 0.4-in/1-cm thick cuts, and remove the seeds.
3. Peel the carrot and julienne it into 2-in/5-cm long slices. Add 2 tablespoons of cooking oil to a skillet and heat it. Then stir-fry the julienned carrot over high heat for 1 minute. Sprinkle a pinch of salt as you stir-fry the carrot.
4. Peel the burdock root and julienne it into 2-in/5-cm long strips. Soak them in cold water for 15 minutes and drain the water. Add 2 tablespoons of cooking oil to a skillet and heat it. Then stir-fry the drained burdock root for 3 to 4 minutes over medium heat. Add all the seasoning ingredients for the burdock root to the skillet and cook everything for 3 more minutes over low heat.

   Tip Make sure to soak the burdock root in cold water before use, as it tastes bitter when used right away.

   Tip When using an onion instead of burdock root, add 1 tablespoon of cooking oil to a skillet, heat it, and stir-fry the julienned onions for 2 minutes over high heat. Then adjust the heat to low, add the seasoning ingredients, and mix well. Continue to stir-fry for 1 minute to remove the moisture.
5. Slice the ham lengthwise into 0.4-in/1-cm thick strips. Add 1 tablespoon of cooking oil to a skillet and heat it. Then evenly cook the sliced ham strips over high heat.
6. Crack the eggs in a bowl with a pinch of salt and beat them well. Add 2 tablespoons of cooking oil to a small pan and heat it. Pour the egg mixture into the pan so that it is 0.4-in/1-cm thick, and cook for 6 to 7 minutes over low heat. Flip the eggs, turn off the heat, and wait for 3 minutes. Cut the cooked eggs into 0.4-in/1-cm wide strips.

   Vegan Leave out the eggs and instead use more cucumbers, burdock root, carrots, or other ingredients.
7. Add all the seasoning ingredients for the rice to the warm rice and mix them well. Lay 1 sheet of *gim* on a rolling mat and spread ½ of the seasoned rice in a thin layer on the *gim*. Spread the rice across ⅘ of the sheet, excluding the top and bottom ends. Then stick a few cooked rice grains in 3 to 4 places along the top edge of the sheet and gently smear them in place to help seal the *gimbap*.

   Tip If you apply a little bit of toasted sesame oil to your hands or your plastic gloves, it will help prevent the rice from sticking to them.

   Tip You can use plastic wrap or parchment paper instead of a rolling mat.

1. 밥을 준비한다. → **218 쪽**
2. 오이는 1cm 두께로 세로로 길게 썰어 씨를 제거한다.
3. 당근은 껍질을 벗겨 5cm 길이로 얇게 채 썬다. 팬에 식용유 2큰술을 두른 뒤 불에 달구고, 채 썬 당근을 강불에서 1분간 볶는다. 볶는 사이사이 소금을 한 꼬집 뿌린다.
4. 우엉은 껍질을 벗겨 5cm 길이로 얇게 채 썬다. 채 썬 우엉은 찬물에 15분간 담가 두었다가 물기를 제거하여 사용한다. 팬에 식용유를 2큰술 두른 뒤 불에 달구고, 물기를 제거한 우엉을 중불에서 3-4분간 볶는다. 팬에 우엉 양념 재료를 모두 넣고 약불에서 3분간 더 볶는다.

   Tip 우엉은 바로 사용하면 쓴맛이 나므로 반드시 찬물에 담가 두었다가 사용한다.

   Tip 우엉 대신 양파를 사용한다면, 팬에 식용유를 1큰술 두른 뒤 불에 달구고, 채 썬 양파를 강불에서 2분간 볶다가 약불로 줄여 양념 재료를 넣고 잘 섞는다. 1분간 더 볶아 수분을 충분히 날린다.
5. 햄은 1cm 두께로 세로로 길게 썬다. 팬에 식용유를 1큰술 두른 뒤 불에 달구고, 길게 썬 햄을 강불에서 고루 굽는다.
6. 볼에 계란과 소금 한 꼬집을 넣어 고루 푼다. 작은 팬에 식용유를 2큰술 두른 뒤 불에 달구고, 계란물을 부어 1cm 두께로 도톰하게 익힌다. 약불에서 6-7분간 익힌 다음 계란을 뒤집고, 불을 끈 뒤 3분간 기다린다. 익은 계란은 1cm 폭으로 길게 썬다.

   Vegan 계란을 제외하고, 오이, 당근, 우엉 등 다른 재료를 넉넉하게 얹는다.
7. 따뜻한 밥에 밥 양념 재료를 모두 넣어 고루 섞는다. 김밥 발 위에 김밥용 김을 한 장 올리고, 김 위에 양념한 밥의 ½ 양을 얹은 뒤 얇게 펼친다. 밥은 김의 위쪽과 아래쪽 끝을 제외하고 안쪽 면적의 ⅘ 정도에만 펼치고, 김밥이 잘 붙도록 김의 위쪽 끝에 밥알을 서너 군데 붙여 뭉개어 둔다.

   Tip 손 혹은 비닐장갑에 참기름을 약간 바르면 밥알이 손에 달라붙지 않는다.

   Tip 김밥 발 대신 비닐랩이나 종이 포일을 사용해도 된다.

8. Arrange 1 cucumber strip, 1 grilled ham strip, 1 egg strip, 1 pickled radish strip, stir-fried carrots, and stir-fried burdock root neatly on the lower ⅓ of the sheet, and roll it upwards starting from the bottom. **photo → page 44** Apply toasted sesame oil on the finished *gimbap* and cut it into bite-sized pieces.
   Tip Adding too many ingredients will make it hard to roll the *gimbap*. Make sure to use the appropriate amount of ingredients in one roll.

PORK BELLY GIMBAP

1. Grill pork belly in a skillet until golden brown, and chop the chili pepper into 0.4-in/1-cm thick pieces.
2. Follow steps 1 to 7.
3. In step 8, place a strip of grilled pork belly on top of the other ingredients and evenly spread *ssamjang* on top of it. Then place the chopped chili pepper on the *ssamjang*, **photo → page 40** and roll the *gimbap* upwards starting from the bottom. **photo → page 44** Apply toasted sesame oil on the finished *gimbap* and cut it into bite-sized pieces.

8. 김의 아래쪽 ⅓지점에 오이 1줄, 구운 햄 1줄, 계란 1줄, 단무지 1줄, 볶은 당근, 볶은 우엉을 가지런히 얹고, 아래쪽부터 위쪽으로 만다. **사진 → 44 쪽** 완성된 김밥에 참기름을 바르고, 먹기 좋은 크기로 썬다.
   Tip 속 재료가 너무 많으면 김밥이 잘 말리지 않으므로, 김밥 한 줄에 넣는 재료의 양을 적절히 조절한다.

삼겹살 김밥

1. 삼겹살을 팬에 노릇하게 굽고, 고추는 1cm 두께로 썬다.
2. 1-7단계를 따라 진행한다.
3. 8단계에서 다른 재료 위에 구운 삼겹살 한 줄을 통째로 올린 뒤 삼겹살에 쌈장을 고루 바른다. 썰어 둔 고추를 쌈장 바른 삼겹살 위에 얹고 **사진 → 40 쪽** 아래쪽부터 위쪽으로 만다. **사진 → 44 쪽** 완성된 김밥에 참기름을 바르고, 먹기 좋은 크기로 썬다.

Gimbap Recipe
Reference Video
김밥 레시피 참고 영상

Special Tip Weverse (2020. 06. 21)

The *gimbap* is less likely to tear apart if you cut it with a blade coated in toasted sesame oil.
칼날에 참기름을 발라 썰면 김밥이 덜 터져요.

Serves **1**
1인분

Difficulty **Low**
난이도 **쉬움**

- BOX LUNCHES JUST FOR YOU -

Toast

j-hope's Egg Sandwich

제이홉표 토스트

STORY → page **46**

INGREDIENTS

- ☐ 2 slices **bread**
- ☐ 2 **eggs**
- ☐ 2 (1 oz / 30 g) strips **bacon**
- ☐ 1½ (0.7 oz / 20 ml) tablespoons **milk**
- ☐ 1⅓ (0.7 oz / 20 g) tablespoons **butter**
- ☐ 1 tablespoon **strawberry jam**
- ☐ ⅓ teaspoon **salt**
- ☐ 2 tablespoons **cooking oil**

- ☐ **식빵** 2개
- ☐ **계란** 2개
- ☐ **베이컨** 2줄 (30g)
- ☐ **우유** 1½큰술 (20ml)
- ☐ **버터** 1⅓큰술 (20g)
- ☐ **딸기잼** 1큰술
- ☐ **소금** ⅓작은술
- ☐ **식용유** 2큰술

HOW TO COOK

1. Crack the eggs in a bowl with some milk and salt and beat them well.
   Vegan If the vegan eggs are watery enough, you can leave out the vegan milk.
2. Add the cooking oil to a skillet and heat it. Pour the egg mixture over medium heat. Adjust the heat to low and make scrambled eggs by stirring the egg mixture with a wooden spoon. Transfer the scrambled eggs to a plate.
3. Wipe the grease off the skillet and grill the bacon on the dry skillet. Transfer the grilled bacon to the plate.
4. Wipe the grease off the skillet and toast 2 slices of bread on the dry skillet until they turn golden brown. Transfer the toasted bread to a plate and spread butter on one side of each slice.
5. Spread strawberry jam on one side of a slice, and add the cooked eggs and bacon on top. Place the other slice of bread on top. **photo → page 51**

1. 볼에 계란과 우유, 소금을 넣고 고루 푼다.
   Vegan 비건 계란이 충분히 묽다면 비건 우유는 제외해도 된다.
2. 팬에 식용유를 두르고 불에 달군 뒤, 중불에서 계란물을 붓는다. 약불로 줄인 후 주걱으로 저어 가며 서서히 익혀 스크램블드에그를 만든다. 스크램블드에그는 접시로 옮겨 둔다.
3. 팬의 기름을 닦아 내고, 마른 팬에 베이컨을 올려 굽는다. 구워진 베이컨은 접시로 옮겨 둔다.
4. 팬의 기름을 닦아 내고, 마른 팬에 식빵 두 장을 노릇하게 굽는다. 구워진 식빵은 접시로 옮기고, 각 식빵의 한쪽 면에 버터를 바른다.
5. 식빵 한 쪽에 딸기잼을 바르고, 익힌 계란과 베이컨을 얹는다. 다른 한 쪽을 포개어 덮는다.
   사진 → 51 쪽

Vegan 비건

- ☐ 2 slices **vegan bread**
- ☐ 3.5 oz / 100 g **vegan eggs**
- ☐ 2 (1 oz / 30 g) strips **vegan bacon**
- ☐ 1½ (0.7 oz / 20 ml) tablespoons **vegan milk**
- ☐ 1⅓ (0.7 oz / 20 g) tablespoons **vegan butter**

- ☐ **비건 식빵** 2장
- ☐ **비건 계란** 100g
- ☐ **비건 베이컨** 2줄 (30g)
- ☐ **비건 우유** 1½큰술 (20ml)
- ☐ **비건 버터** 1⅓큰술 (20g)

Gluten-free 글루텐 프리

- ☐ 2 slices **gluten-free bread**
- ☐ 1 tablespoon **gluten-free strawberry jam**

- ☐ **글루텐 프리 식빵** 2장
- ☐ **글루텐 프리 딸기잼** 1큰술

Special Tip

You can use blueberry or apple jam instead of strawberry jam.
But it tastes best with strawberry jam!

딸기잼 대신 블루베리잼, 사과잼 등을 사용해도 되지만 딸기잼을 사용했을 때 가장 맛있어요!

[VLOG] j-hope (2019. 10. 03)

Serves **1**
1인분

Difficulty **Low**
난이도 **쉬움**

- BOX LUNCHES JUST FOR YOU -

Toast

Jung Kook's Egg Sandwich

정국표 토스트

STORY ⟶ page **46**

INGREDIENTS

- ☐ 2 slices **bread**
- ☐ 2 **eggs**
- ☐ 1 slice **cheese**
- ☐ 1⅓ (0.7 oz / 20 g) tablespoons **butter**
- ☐ 1 tablespoon **ketchup**
- ☐ ⅓ teaspoon **salt**

- ☐ **식빵** 2장
- ☐ **계란** 2개
- ☐ **슬라이스 치즈** 1장
- ☐ **버터** 1⅓큰술 (20g)
- ☐ **케첩** 1큰술
- ☐ **소금** ⅓작은술

HOW TO COOK

1. Crack the eggs in a bowl with some salt and beat them well.
2. Melt the butter in a wide skillet and spread the egg mixture widely on medium heat. Then place 2 slices of bread side by side on top of the egg mixture.
3. When the egg mixture turns opaque, adjust the heat to low and fold the eggs lying beside the bread onto the bread slices.
4. Place the sliced cheese on one side and spread ketchup on the other side. Fold the two slices of bread together. **photo → page 49**

1. 볼에 계란과 소금을 넣고 고루 푼다.
2. 넓은 팬에 버터를 녹인 뒤, 중불에서 계란물을 부어 넓게 펼친다. 그 위에 바로 식빵 2장을 나란히 얹는다.
3. 계란물이 익어 불투명해지면 약불로 줄인 후 식빵 바깥쪽의 계란을 식빵 위로 접어 올린다.
4. 식빵 한 쪽에 슬라이스 치즈를 올리고, 다른 한 쪽에는 케첩을 바른다. 두 쪽이 포개어지도록 접는다. **사진 → 49 쪽**

Vegan 비건

- ☐ 2 slices **vegan bread**
- ☐ 3.5 oz / 100 g **vegan eggs**
- ☐ 1 slice **vegan cheese**
- ☐ 1⅓ (0.7 oz / 20 g) tablespoons **vegan butter**

- ☐ **비건 식빵** 2장
- ☐ **비건 계란** 100g
- ☐ **비건 슬라이스 치즈** 1장
- ☐ **비건 버터** 1⅓큰술 (20g)

Gluten-free 글루텐 프리

- ☐ 2 slices **gluten-free bread**
- ☐ 1 tablespoon **gluten-free ketchup**

- ☐ **글루텐 프리 식빵** 2장
- ☐ **글루텐 프리 케첩** 1큰술

3

Slurp! Noodle Dishes

후루룩 면 요리

Kalguksu 칼국수
Jjajang Ramyeon 짜장 라면

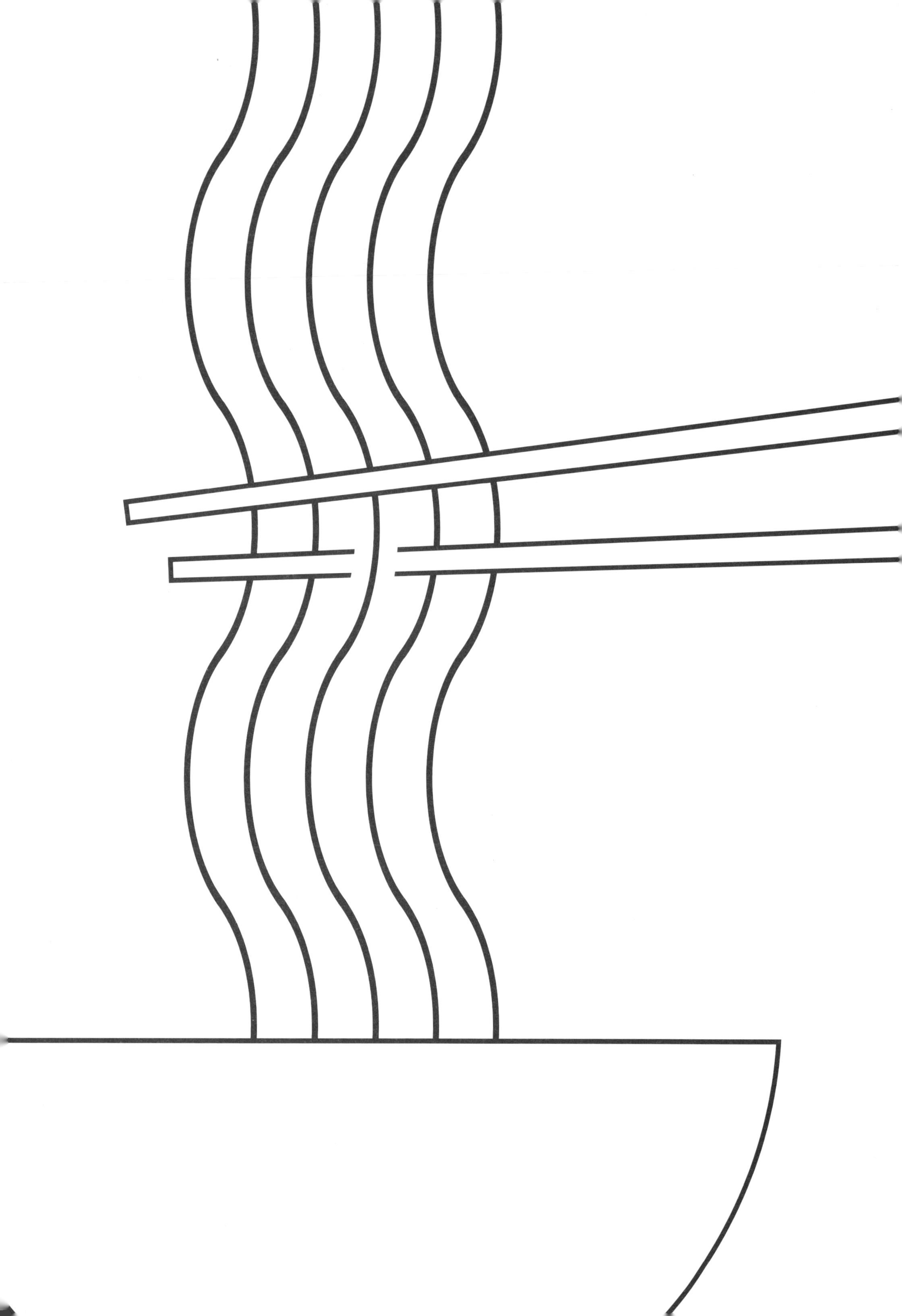

Serves **2** to **3**
2-3인분

Difficulty **Mid**
난이도 **보통**

- SLURP! NOODLE DISHES -

Kalguksu

Noodle Soup

칼국수

STORY ⟶ page **56**

INGREDIENTS

- ☐ 10.5 oz / 300 g clams
- ☐ 7 oz / 200 g mussels
- ☐ 7 oz / 200 g scallops
- ☐ ¼ (3.5 oz / 100 g) zucchini
- ☐ ⅓ (2.5 oz / 70 g) onion
- ☐ ⅙ (2 oz / 60 g) carrot
- ☐ ¼ (4 in / 10 cm, 1 oz / 30 g) large green onion
- ☐ 2 tablespoons fermented anchovy sauce
- ☐ 1⅓ tablespoons salt
- ☐ 1 tablespoon minced garlic
- ☐ ½ teaspoon black pepper

- ☐ **조개** 300g
- ☐ **홍합** 200g
- ☐ **가리비** 200g
- ☐ **주키니** ¼개 (100g)
- ☐ **양파** ⅓개 (70g)
- ☐ **당근** ⅙개 (60g)
- ☐ **대파** ¼대 (10cm, 30g)
- ☐ **멸치 액젓** 2큰술
- ☐ **소금** 1⅓큰술
- ☐ **다진 마늘** 1큰술
- ☐ **후추** ½작은술

NOODLES (10.5 oz / 300 g)

- ☐ 3 cups (14 oz / 400 g) all-purpose flour
- ☐ ⅔ cup (4.7 oz / 140 ml) lukewarm water
- ☐ ½ teaspoon cooking oil
- ☐ ⅓ teaspoon salt

면 (300g)

- ☐ **밀가루(중력분)** 3컵 (400g)
- ☐ **미지근한 물** ⅔컵 (140ml)
- ☐ **식용유** ½작은술
- ☐ **소금** ⅓작은술

ANCHOVY BROTH 5 cups (34 oz / 1 L)

- ☐ 40 (1.5 oz / 40 g) dried anchovies for stock
- ☐ 1 sheet (2 in × 3 in / 5 cm × 7 cm) dried kelp
- ☐ 6 cups (40.5 oz / 1.2 L) water

멸칫국물 5컵 (1L)

- ☐ **국물용 멸치** 40마리 (40g)
- ☐ **건조 다시마** 1장 (5cm × 7cm)
- ☐ **물** 6컵 (1.2L)

Vegan 비건

- ☐ 2 oz / 60 g mushrooms

- ☐ **버섯** 60g

Gluten-free 글루텐 프리

- ☐ 2 tablespoons gluten-free fermented anchovy sauce
- ☐ 2 tablespoons gluten-free soy sauce (for seasoning sauce)

- ☐ **글루텐 프리 멸치 액젓** 2큰술
- ☐ **글루텐 프리 간장** 2큰술 (양념장)

GLUTEN-FREE NOODLES (10.5 oz / 300 g)

- ☐ 3 cups (14 oz / 400 g) rice flour for noodles
- ☐ ⅔ cup (4.7 oz / 140 ml) of warm water around 176°F / 80°C
- ☐ ½ teaspoon cooking oil
- ☐ ½ teaspoon salt

글루텐 프리 면 (300g)

- ☐ **쌀가루(제면용)** 3컵 (400g)
- ☐ **80°C 정도의 따뜻한 물** ⅔컵 (140ml)
- ☐ **식용유** ½작은술
- ☐ **소금** ½작은술

SEASONING SAUCE

- ☐ 1 (0.3 oz / 10 g) hot chili pepper
- ☐ 2 (0.3 oz / 10 g) scallions
- ☐ 2 tablespoons soy sauce
- ☐ 1 teaspoon sugar
- ☐ 1 teaspoon toasted sesame oil
- ☐ 1 teaspoon toasted sesame seeds
- ☐ ½ teaspoon *gochutgaru* (Korean chili powder)

양념장

- ☐ **청양고추** 1개 (10g)
- ☐ **쪽파** 2대 (10g)
- ☐ **간장** 2큰술
- ☐ **설탕** 1작은술
- ☐ **참기름** 1작은술
- ☐ **참깨** 1작은술
- ☐ **고춧가루** ½작은술

Tips

- You can use linguine pasta noodles (7 oz / 200 g) instead of *kalguksu* noodles. Add the pasta noodles to boiling water with 1 tablespoon of salt and cook for 10 minutes. Then place them in a strainer before putting them into the broth.
- You can use the same amount of Korean zucchini instead of regular zucchini.
- You can use the same amount of fish sauce instead of fermented anchovy sauce.
- You can use the same amount of chives instead of scallions (for seasoning sauce).

- 칼국수면 대신 파스타 면(링귀니) 200g 을 사용해도 된다. 파스타 면은 끓는 물에 소금 1큰술을 넣고 10분간 삶아 체에 밭쳐 두었다가 칼국수 국물에 넣는다.
- 주키니 대신 같은 분량의 애호박을 사용해도 된다.
- 멸치 액젓 대신 같은 분량의 피쉬 소스를 사용해도 된다.
- 쪽파 대신 같은 분량의 부추를 사용해도 된다. (양념장)

HOW TO COOK

1. Scrub the clams, mussels, and scallops under running water. Once they are clean, put them in a large bowl and submerge them in water. Dissolve 1 tablespoon of salt in the water and leave them in a dark place for about 1 hour to remove sediment.
   Tip When washing shellfish, rub the shells against each other to remove any sediment stuck on the surface. Be careful not to cut your hands on any broken shells while washing them. Wearing rubber gloves or cotton work gloves is a good idea.
   Vegan Leave out all the seafood.

1. 조개, 홍합, 가리비를 흐르는 물에 비벼 씻는다. 큰 볼에 깨끗하게 씻은 조개, 홍합, 가리비를 모두 넣고, 재료가 잠길 정도의 물을 붓는다. 물에 소금 1큰술을 넣어 녹인 뒤, 어두운 환경에서 1시간 정도 해감한다.
   Tip 조개류를 씻을 때는 껍데기끼리 문질러 겉에 붙은 이물질을 제거한다. 씻는 동안 깨진 껍데기에 손이 베지 않도록 주의한다. 고무장갑이나 목장갑을 끼고 씻어도 좋다.
   Vegan 해산물은 모두 제외한다.

PREPARING NOODLES (steps 2 to 5)

2. Combine all the ingredients for the noodles in a bowl, mix well, and knead to form a dough. Once complete, cover the dough with plastic wrap and let it rest in the refrigerator for 15 minutes. Then take the fermented dough out of the refrigerator and let it come to room temperature.
   Tip The condition of the dough may differ depending on the temperature and humidity. Add water to the dough in small amounts until you reach the perfect consistency.
   Gluten-free Use rice flour instead of wheat flour. Sift the rice flour before using it. Put the rice flour and salt in a bowl, pour in half of the warm water, and mix well. Then add the remaining water and cooking oil, and knead for 10 minutes. Cover the rice flour dough with plastic wrap and let it rest in the refrigerator for 30 minutes.
3. Sprinkle flour evenly on a cutting board and a rolling pin. Then roll out the dough as thin as possible using the rolling pin. Sprinkle flour on the pressed dough and fold it in half. Sprinkle flour again and fold the dough in half in the same direction, and then repeat this one more time so that the dough is folded into a shape of a long bar. Starting from one end, thinly slice the dough into 0.2-in/0.5-cm wide strips. **photo → page 59**
4. Sprinkle flour on the strips of dough to prevent them from sticking to each other. Loosen the strips by hand. Sprinkle flour on the completed noodles and set them aside.
5. Cut the zucchini into 0.2-in/0.5-cm thick slices, and julienne the onion and carrot into 0.1-in/0.3-cm thick strips. Slice the large green onion diagonally into 0.2-in/0.5-cm thick pieces.
   Vegan Cut the mushrooms into bite-sized pieces, too.

면 준비하기 (2-5단계)

2. 볼에 면 재료를 모두 넣고 잘 섞어 치댄다. 완성된 반죽은 비닐을 씌워 냉장고에서 15분간 숙성시킨다. 숙성시킨 반죽은 실온에 잠시 두어 냉기가 빠지도록 한다.
   Tip 반죽의 상태는 온도나 습도에 따라 달라질 수 있는데, 물을 조금씩 나누어 넣으며 적당히 찰기 있는 반죽이 되도록 조절한다.
   Gluten-free 밀가루 대신 쌀가루를 사용한다. 쌀가루는 체에 한 번 내려 사용한다. 볼에 쌀가루와 소금을 넣고 따뜻한 물을 반만 부어 섞은 다음, 나머지 물과 식용유를 넣어 10분간 치댄다. 완성된 쌀가루 반죽은 비닐을 씌워 냉장고에서 30분간 숙성시켜 사용한다.
3. 도마와 밀대에 밀가루를 고르게 뿌리고, 반죽을 밀대로 밀어 최대한 얇게 펼친다. 펼쳐진 반죽 위에 밀가루를 뿌리고, 반으로 접는다. 그 위에 다시 밀가루를 뿌리고 같은 방향으로 반으로 접은 다음, 같은 동작을 한번 더 반복한다. 긴 막대 모양으로 접힌 반죽을 한쪽 끝에서부터 0.5cm 두께로 얇게 썬다. **사진 → 59 쪽**
4. 썰어 둔 반죽이 서로 달라붙지 않도록 밀가루를 뿌린 다음, 한 도막씩 손으로 푼다. 완성된 면에 밀가루를 뿌려 둔다.
5. 주키니는 0.5cm 두께로 썰고, 양파와 당근은 0.3cm 두께로 채 썬다. 대파는 0.5cm 두께로 어슷하게 썬다.
   Vegan 버섯도 먹기 좋은 크기로 썬다.

PREPARING BROTH (steps 6 to 7)

6. Prepare the anchovy broth. → **page 220**

   Vegan Prepare the vegetable broth instead of the anchovy broth.
   → **page 219**

7. Drain the clean clams, mussels, and scallops. Put them in a pot, add 5 cups of anchovy broth, and boil them for 20 minutes. Then, take the clams, mussels, and scallops out of the broth. Separate the scallops from their shells and cut them into bite-sized pieces. Strain the broth.

   Tip The umami flavor comes from the clam shells, so make sure you boil the clams with the shells when you make the broth.

8. Pour the strained broth into a pot and boil it again. When the broth starts boiling, add the noodles, prepared vegetables, minced garlic, and fermented anchovy sauce, and let it simmer for 10 more minutes over medium heat.

   Tip When using pasta noodles instead of *kalguksu* noodles, use pre-cooked pasta noodles.

   Vegan Leave out the fermented anchovy sauce. Add the prepared vegetables and mushrooms, and simmer for 10 more minutes over medium heat.

9. When the edges of the *kalguksu* noodles become translucent, add the clams, mussels, and sliced scallops, and simmer them for 1 to 2 more minutes. Season with salt and pepper to taste.

   Tip Season with salt at the very last step, as the clams add saltiness. It's okay if it tastes a bit bland, as you can add the seasoning sauce as you eat.

10. Finely chop the hot chili pepper and put it in a bowl with the rest of the seasoning sauce ingredients and mix well. Add them to your *kalguksu* in small amounts to taste.

국물 준비하기 (6-7단계)

6. 멸칫국물을 준비한다. → **220 쪽**

   Vegan 멸칫국물 대신 채소 국물을 준비한다.
   → **219 쪽**

7. 해감한 조개, 홍합, 가리비의 물기를 제거한 뒤 냄비에 모두 담고 멸칫국물 5컵을 부어 20분간 끓인다. 그다음, 조개, 홍합, 가리비를 건져 내고 가리비는 살만 분리하여 먹기 좋은 크기로 썬다. 국물은 체에 거른다.

   Tip 조개 껍데기에서 감칠맛이 우러나므로 국물을 우릴 때는 조개를 껍질째 넣고 끓인다.

8. 체에 거른 국물을 냄비에 부어 다시 끓인다. 국물이 끓어오르면 면과 손질한 채소, 다진 마늘, 멸치 액젓을 넣고 중불에서 10분간 더 끓인다.

   Tip 칼국수면 대신 파스타 면 사용 시, 미리 익혀 둔 파스타 면을 넣는다.

   Vegan 멸치 액젓은 제외하고, 손질한 채소와 버섯을 함께 넣고 중불에서 10분간 더 끓인다.

9. 칼국수면의 가장자리가 투명해지면 조개, 홍합과 썰어 둔 가리비 살을 넣고 1-2분간 더 끓인다. 소금과 후추를 넣어 간을 맞춘다.

   Tip 조개에 염분이 있기 때문에 소금 간은 마지막에 한다. 양념장을 곁들일 수 있으므로 간은 약간 싱겁게 맞춰도 된다.

10. 청양고추를 잘게 다진 뒤, 양념장 재료와 함께 볼에 넣어 고루 섞는다. 기호에 따라 칼국수에 조금씩 넣어 먹는다.

Serves **2**
2인분

Difficulty **Low**
난이도 **쉬움**

- SLURP! NOODLE DISHES -

Jjajang Ramyeon

Instant Noodles in Black Bean Sauce

짜장 라면

STORY ⟶ page **64**

INGREDIENTS

- ☐ 2 packages **instant *jjajang ramyeon* (instant noodles in black bean sauce)**
- ☐ 7 oz / 200 g **boneless beef ribs**
- ☐ ⅓ (2.5 oz / 70 g) **onion**
- ☐ 1½ cups (10 oz / 300 ml) **water**
- ☐ ⅓ teaspoon **salt**
- ☐ 2 tablespoons **cooking oil**

- ☐ **인스턴트 짜장 라면** 2개
- ☐ **소고기 갈빗살** 200g
- ☐ **양파** ⅓개 (70g)
- ☐ **물** 1½컵 (300ml)
- ☐ **소금** ⅓작은술
- ☐ **식용유** 2큰술

Tips

- For beef, you can use the same amount of sirloin or tenderloin instead of boneless ribs.
- You can use 1.5 oz / 40 g of scallions or chives instead of onion.

- 소고기는 갈빗살 대신 같은 분량의 구이용 등심이나 안심을 사용해도 된다.
- 양파 대신 쪽파나 부추 40g을 사용해도 된다.

Vegan 비건

- ☐ 2 packages **vegan instant *jjajang ramyeon***
- ☐ ½ (3.5 oz / 100 g) **onion**
- ☐ 2 (1.5 oz / 40 g) **button mushrooms**
- ☐ 1 teaspoon (0.2 oz / 5 g) **vegan butter**

- ☐ **비건 인스턴트 짜장 라면** 2개
- ☐ **양파** ½개 (100g)
- ☐ **양송이버섯** 2개 (40g)
- ☐ **비건 버터** 1작은술 (5g)

Gluten-free 글루텐 프리

- ☐ 2 packages **gluten-free instant *jjajang ramyeon***

- ☐ **글루텐 프리 인스턴트 짜장 라면** 2개

HOW TO COOK

1. Pour enough water to cook the noodles into a pot and bring it to a boil. When the water starts boiling, add the noodles and cook for the time indicated on the package. (Do not add the seasoning powder or flakes yet.) Place the cooked noodles in a strainer and drain them.
2. Slice the onion into 0.8 in × 1.2 in / 2 cm × 3 cm squares.
3. Add 1 tablespoon of cooking oil to a skillet and heat it. Grill the beef over high heat and sprinkle some salt on it as you grill. Grill for 1 to 2 minutes until both sides are golden brown, then adjust the heat to low and cook for 3 to 4 minutes.
   Vegan Leave out the beef.
4. Remove the beef from the skillet. Add 1 tablespoon of cooking oil to the leftover oil in the same skillet. Add the sliced onion and stir-fry for 5 minutes over medium heat. Then add 1½ cups (10 oz / 300 ml) of water, along with the seasoning powder and flakes from the instant noodles, and stir well. Simmer for 2 minutes over low heat.
   Vegan Heat a skillet and melt the vegan butter. Add the sliced onion and quartered button mushrooms and stir-fry for 3 minutes over medium heat. Add the water, seasoning powder, and flakes in the same skillet, and stir well. Simmer for 2 minutes over low heat.
5. Put the boiled noodles in the skillet and mix well. **photo → page 70**
   Transfer the finished *jjajang ramyeon* to a bowl. Cut the grilled beef into bite-sized pieces and place them on the noodles.

1. 면을 익히기 위해 냄비에 물을 적당량 붓고 끓인다. 물이 끓으면 면을 넣어 포장지에 적힌 시간만큼 익힌다. (분말수프와 건더기 수프는 아직 넣지 않는다.) 익은 면은 체에 밭쳐 물기를 제거한다.
2. 양파는 2cm × 3cm 크기로 네모나게 썬다.
3. 팬에 식용유 1큰술을 두르고 불에 달군 뒤, 강불에서 소고기를 굽는다. 앞뒷면이 노릇해질 정도로 1-2분간 구운 뒤, 약불로 줄여 3-4분간 고기를 익힌다. 굽는 사이사이 소금을 뿌린다.
   Vegan 소고기는 제외한다.
4. 팬에서 고기를 건져 내고, 기름이 남아 있는 팬에 식용유를 1큰술 추가한다. 팬에 썰어 둔 양파를 넣고 중불에서 5분간 볶는다. 물 1½컵(300ml)과 분말수프, 건더기 수프를 모두 넣고 잘 저은 다음 약불에서 2분간 끓인다.
   Vegan 팬을 달구고 비건 버터를 녹인다. 썰어 둔 양파와 4등분한 양송이버섯을 넣고 중불에서 3분간 볶는다. 팬에 물과 분말수프, 건더기 수프를 모두 넣고 잘 저은 다음 약불에서 2분간 끓인다.
5. 삶아둔 면을 팬에 넣어 고루 버무린다.
   사진 → 70쪽 완성된 짜장 라면을 그릇에 옮겨 담고 구운 고기를 먹기 좋은 크기로 잘라 얹는다.

Special Tip <Run BTS!> EP. 131

An over-easy fried egg makes a good topping for *jjajang ramyeon*.
짜장 라면에 반숙으로 조리한 계란프라이를 올려 먹어도 잘 어울려요.

4

Spicy Meat Dishes

매콤한 고기 요리

Dakgalbi 닭갈비

Jeyuk Bokkeum 제육볶음

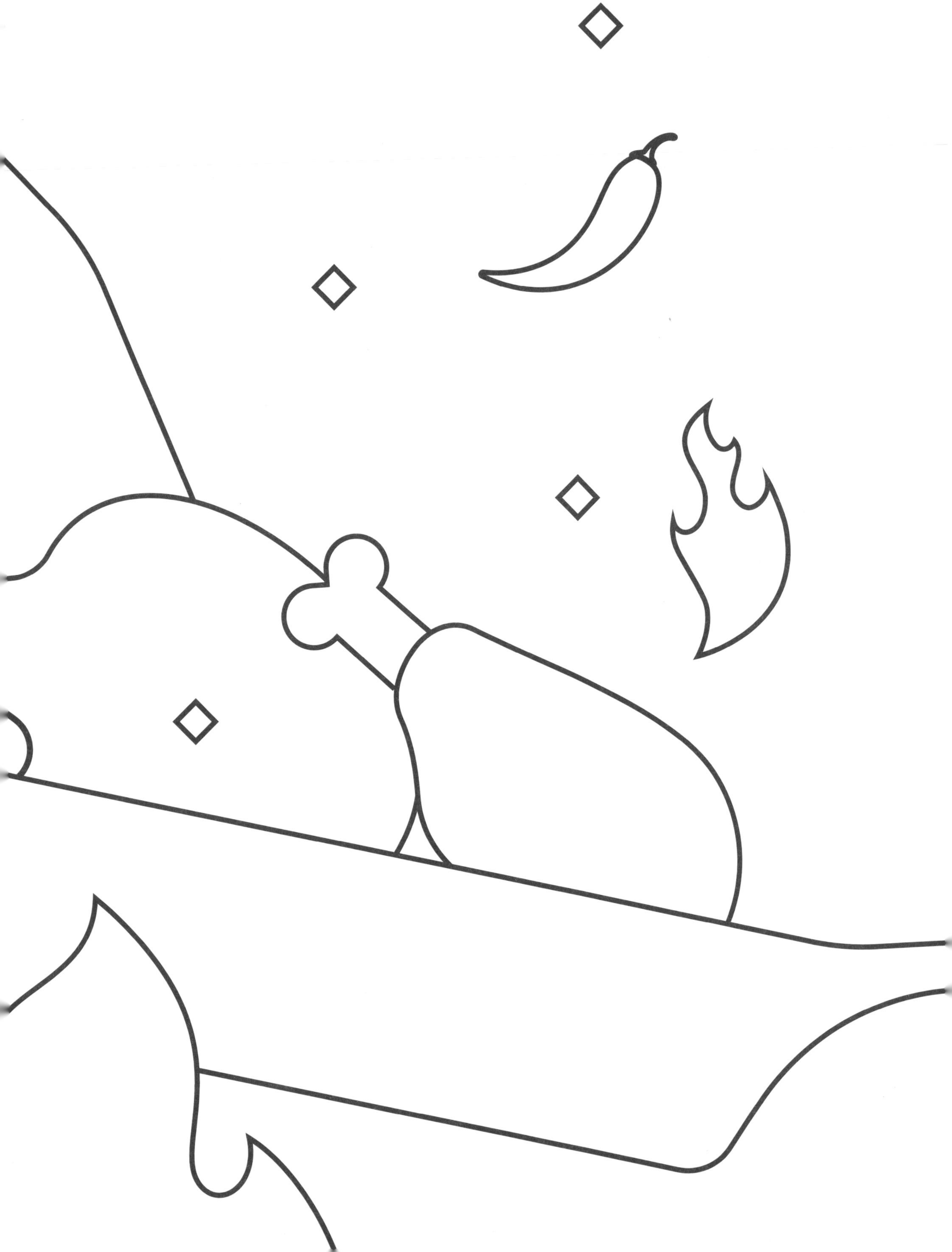

Serves **2**
2인분

Difficulty **Mid**
난이도 **보통**

Dakgalbi

Spicy Stir-Fried Chicken

닭갈비

STORY ⟶ page **74**

INGREDIENTS

- ☐ 14 oz / 400 g boneless chicken thighs
- ☐ 12 (3.5 oz / 100 g) bite-sized rice cakes
- ☐ ¼ (3.5 oz / 100 g) cabbage
- ☐ 1 (3 oz / 80 g) sweet potato or potato
- ☐ ⅓ (2.5 oz / 70 g) onion
- ☐ ½ (8 in / 20 cm, 2 oz / 60 g) large green onion
- ☐ 1 (0.3 oz / 10 g) hot chili pepper
- ☐ 1 (0.3 oz / 10 g) red chili pepper
- ☐ 5 (0.3 oz / 10 g) perilla leaves (*kkaennip*)
- ☐ 2 tablespoons *cheongju* (refined rice wine)
- ☐ 2 tablespoons cooking oil

- ☐ **닭고기 다리 살** 400g
- ☐ **떡볶이 떡** 12개 (100g)
- ☐ **양배추** ¼개 (100g)
- ☐ **고구마** 또는 **감자** 1개 (80g)
- ☐ **양파** ⅓개 (70g)
- ☐ **대파** ½대 (20cm, 60g)
- ☐ **청양고추** 1개 (10g)
- ☐ **홍고추** 1개 (10g)
- ☐ **깻잎** 5장 (10g)
- ☐ **청주** 2큰술
- ☐ **식용유** 2큰술

GOCHUJANG SEASONING

- ☐ 3 tablespoons soy sauce
- ☐ 2 tablespoons *gochujang* (Korean red chili paste)
- ☐ 2 tablespoons toasted sesame oil
- ☐ 1 tablespoon honey
- ☐ 1 tablespoon *cheongju*
- ☐ 1 tablespoon sugar
- ☐ 1 tablespoon minced garlic
- ☐ ½ teaspoon minced ginger
- ☐ ½ teaspoon black pepper

고추장 양념

- ☐ **간장** 3큰술
- ☐ **고추장** 2큰술
- ☐ **참기름** 2큰술
- ☐ **꿀** 1큰술
- ☐ **청주** 1큰술
- ☐ **설탕** 1큰술
- ☐ **다진 마늘** 1큰술
- ☐ **다진 생강** ½작은술
- ☐ **후추** ½작은술

Gluten-free 글루텐 프리

- ☐ 12 (3.5 oz / 100 g) gluten-free bite-sized rice cakes
- ☐ 3 tablespoons gluten-free soy sauce
- ☐ 2 tablespoons gluten-free *gochujang*
- ☐ 1 tablespoon gluten-free oyster sauce (for soy sauce seasoning)

- ☐ **글루텐 프리 떡볶이 떡** 12개 (100g)
- ☐ **글루텐 프리 간장** 3큰술
- ☐ **글루텐 프리 고추장** 2큰술
- ☐ **글루텐 프리 굴소스** 1큰술 (간장 양념)

SOY SAUCE SEASONING

- ☐ 3 tablespoons soy sauce
- ☐ 2 tablespoons toasted sesame oil
- ☐ 1 tablespoon oyster sauce
- ☐ 1 tablespoon honey
- ☐ 1 tablespoon *cheongju*
- ☐ 1 tablespoon sugar
- ☐ 1 tablespoon minced garlic
- ☐ ½ teaspoon minced ginger
- ☐ ½ teaspoon black pepper

간장 양념

- ☐ **간장** 3큰술
- ☐ **참기름** 2큰술
- ☐ **굴소스** 1큰술
- ☐ **꿀** 1큰술
- ☐ **청주** 1큰술
- ☐ **설탕** 1큰술
- ☐ **다진 마늘** 1큰술
- ☐ **다진 생강** ½작은술
- ☐ **후추** ½작은술

* If you want to make non-spicy soy sauce *dakgalbi*, use the soy sauce seasoning ingredients and leave out the hot chili pepper.

* 맵지 않은 간장 닭갈비를 만들고 싶다면, 간장 양념 재료를 사용하고 청양고추는 제외한다.

Tips

- You can use other chicken parts instead of chicken thighs. Boneless chicken is easier to cook.
- You can use the same amount of pasta shells (called conchiglie) instead of rice cakes.
- You can use peperoncino instead of hot chili pepper.
- You can use the same amount of oligo syrup or *mullyeot* (corn syrup) instead of honey.
- You can use soju or dry white wine instead of *cheongju*.
- Perilla leaves add a fragrant aroma, but you can leave them out if they are unavailable.
- This dish goes well with mozzarella cheese, *ramyeon* noodles (instant noodles), and glass noodles (*dangmyeon*), so feel free to add them based on your preference.

- 닭고기는 다리 살 대신 다른 부위를 사용해도 된다. 순살로 준비해야 조리하기 쉽다.
- 떡 대신 같은 분량의 파스타 셸(콘킬리에)을 사용해도 잘 어울린다.
- 청양고추 대신 페페론치노를 사용해도 된다.
- 꿀 대신 같은 양의 올리고당이나 물엿을 사용해도 된다.
- 청주 대신 소주 또는 달지 않은 화이트 와인을 사용해도 된다.
- 깻잎은 향긋함을 더해 주는데, 구하기 어렵다면 제외해도 된다.
- 취향에 따라 모차렐라 치즈, 라면 사리, 당면 등을 추가해도 잘 어울린다.

HOW TO COOK

1. Remove the small bones from the chicken and cut into 1.2-in/3-cm wide pieces. Put the chicken pieces in a bowl, pour in the *cheongju*, and mix well.
2. Put all the ingredients for the seasoning in another bowl and mix well. Mix half of the seasoning with the prepared chicken. Place the seasoned chicken and the leftover seasoning in the refrigerator to marinate for at least 30 minutes.
3. Soak the bite-sized rice cakes in water for 10 minutes.
4. Cut the sweet potato or potato into flat 0.4-in/1-cm thick slices. Thinly slice the onion into 0.2-in/0.5-cm thick slices, and cut the cabbage and perilla leaves into 0.7-in/2-cm thick slices. Cut the large green onion, hot chili pepper, and red chili pepper diagonally into 0.1-in/0.2-cm thick slices.
5. Add the cooking oil to a skillet and heat it. Then stir-fry the seasoned chicken, soaked bite-sized rice cakes, sliced sweet potato or potato, onion, and cabbage over high heat. After stir-frying for about 5 to 6 minutes, add the leftover seasoning, sliced large green onion, hot chili pepper, and red pepper and continue to stir-fry for another 7 to 8 minutes over medium heat. (Make sure not to add the perilla leaves yet.)
6. When the chicken is fully cooked, add the perilla leaves and mix all the ingredients together.

   Tip The dish is ready to eat once the cabbage becomes transparent and you can easily poke through the chicken with chopsticks or a fork.

   Tip After eating most of the *dakgalbi*, stir-fry rice in the leftover sauce to make *dakgalbi* fried rice. Adding some seaweed flakes and toasted sesame oil enhances the savory flavor, making it more delicious.

1. 닭고기는 잔뼈를 발라낸 뒤 3cm 폭으로 썰고 볼에 넣어 청주와 함께 고루 섞어 둔다.
2. 다른 볼에 양념 재료를 모두 넣고 고루 섞는다. 양념의 반을 덜어 손질한 닭고기와 함께 버무린 뒤, 양념한 고기와 남은 양념을 모두 냉장고에 넣어 30분 이상 숙성시킨다.
3. 떡볶이 떡은 물에 담가 10분간 불린다.
4. 고구마 또는 감자는 1cm 두께로 납작하게 썬다. 양파는 0.5cm 두께로 채 썰고, 양배추와 깻잎은 2cm 두께로 썬다. 대파, 청양고추, 홍고추는 0.2cm 두께로 어슷하게 썬다.
5. 팬에 식용유를 두르고 달군 뒤, 양념한 닭고기와 불린 떡볶이 떡, 썰어 둔 고구마나 감자, 양파, 양배추를 넣어 강불에서 볶는다. 5-6분간 볶다가 남은 양념과 썰어 둔 대파, 청양고추, 홍고추를 넣고 중불에서 7-8분간 더 볶는다. (깻잎은 아직 넣지 않는다.)
6. 닭고기가 완전히 익으면 깻잎을 넣고 모든 재료를 고루 섞는다.

   Tip 양배추가 투명해지고 닭고기를 젓가락이나 포크로 찔렀을 때 잘 들어가면 먹어도 좋다.

   Tip 닭갈비를 어느 정도 먹고 나서 남은 양념에 밥을 넣어 볶으면 닭갈비 볶음밥이 완성된다. 볶음밥을 만들 때 김 가루와 참기름을 적당량 추가하면 감칠맛이 더해져 더욱 맛있다.

Special Tip <IN THE SOOP BTS ver. Season 2> EP. 5

If you place the cabbage, onion, and other vegetables toward the bottom of the skillet and cook them first, it will be easier to stir-fry, as the vegetables will lose their volume quickly.

양배추와 양파 등의 채소를 팬 바닥쪽으로 넣어 익히면 숨이 빠르게 죽어 재료를 볶기 쉬워져요.

MEMO

Serves **2 to 3**
2-3인분

Difficulty **Low**
난이도 쉬움

Jeyuk Bokkeum

Spicy Stir-Fried Pork

제육볶음

STORY → page **82**

INGREDIENTS

- ☐ 1.3 lb / 600 g pork belly (*samgyeopsal*)
- ☐ ½ (3.5 oz / 100 g) onion
- ☐ ½ (8 in / 20 cm, 2 oz / 60 g) large green onion
- ☐ 1 tablespoon toasted sesame seeds
- ☐ 2 tablespoons cooking oil

- ☐ **돼지고기 삼겹살** 600g
- ☐ **양파** ½개 (100g)
- ☐ **대파** ½대 (20cm, 60g)
- ☐ **참깨** 1큰술
- ☐ **식용유** 2큰술

SEASONING

- ☐ 3 tablespoons *gochujang* (Korean red chili paste)
- ☐ 3 tablespoon *cheongju* (refined rice wine)
- ☐ 1 tablespoon honey
- ☐ 1 tablespoon soy sauce
- ☐ 1 tablespoon sugar
- ☐ 1 tablespoon *gochutgaru* (Korean chili powder)
- ☐ 1 tablespoon toasted sesame oil
- ☐ ½ teaspoon minced garlic
- ☐ ¼ teaspoon ginger powder
- ☐ a dash of black pepper

양념

- ☐ **고추장** 3큰술
- ☐ **청주** 3큰술
- ☐ **꿀** 1큰술
- ☐ **간장** 1큰술
- ☐ **설탕** 1큰술
- ☐ **고춧가루** 1큰술
- ☐ **참기름** 1큰술
- ☐ **다진 마늘** ½작은술
- ☐ **생강 가루** ¼작은술
- ☐ **후추** 약간

Gluten-free 글루텐 프리

- ☐ 3 tablespoons gluten-free *gochujang*
- ☐ 1 tablespoon gluten-free soy sauce

- ☐ **글루텐 프리 고추장** 3큰술
- ☐ **글루텐 프리 간장** 1큰술

Tips

- You can replace the pork belly with pork shoulder or front hock.
- You can use soju or dry white wine instead of *cheongju*.

- 돼지고기는 삼겹살 대신 목살 또는 앞다리 살을 사용해도 된다.
- 청주 대신 소주 또는 달지 않은 화이트 와인을 사용해도 된다.

HOW TO COOK

1. Cut the pork belly into 1.5-in/4-cm wide pieces.
2. Put all the ingredients for the seasoning in a bowl and mix them well. Mix half of the seasoning with the sliced pork. Place the seasoned pork and the leftover seasoning in the refrigerator to marinate for at least 15 minutes.
3. Slice the large green onion into 2-in/5-cm long pieces, and thinly slice the onion into 0.5-in/1-cm thick pieces.
4. Add the cooking oil to a skillet and heat it. Cook the marinated pork for 2 to 3 minutes over medium heat. Flip the meat frequently to prevent the seasoning from burning.
5. Add the remaining marinated seasoning along with the sliced large green onion and onion into the skillet and stir-fry for 5 to 6 minutes over medium heat.
6. Transfer the *jeyuk bokkeum* to a plate and decorate with toasted sesame seeds.

1. 삼겹살은 4cm 폭으로 썬다.
2. 볼에 양념 재료를 모두 넣고 고루 섞는다. 양념의 반을 덜어 잘라 둔 돼지고기와 함께 버무린 뒤, 양념한 돼지고기와 남은 양념을 모두 냉장고에 넣어 15분 이상 숙성시킨다.
3. 대파는 5cm 길이로 썰고, 양파는 1cm 두께로 채 썬다.
4. 팬에 식용유를 두르고 불에 달군 뒤, 양념하여 숙성시킨 돼지고기를 중불에서 2-3분간 익힌다. 이때, 양념이 타지 않도록 자주 뒤집는다.
5. 남은 양념과 썰어 둔 대파, 양파를 모두 팬에 넣고 중불에서 5-6분간 더 볶는다.
6. 접시에 옮겨 담고 참깨를 뿌려 장식한다.

Special Tip <Run BTS!> EP. 102

If you are not good at eating spicy foods, try this dish with some cooked rice.
매운 음식을 잘 못 먹는다면 밥을 곁들여 보세요.

5

Dishes for Special Feasts

특별한 만찬

Miyeokguk 미역국
Japchae 잡채
Tteokguk 떡국
Bulgogi 불고기

Serves **2**
2인분

Difficulty **Mid**
난이도 **보통**

DISHES FOR SPECIAL FEASTS

BIRTHDAY

Miyeokguk

Seaweed Soup

미역국

STORY → page **92**

INGREDIENTS

- ☐ 0.7 oz / 20 g **dried seaweed** or 5.3 oz / 150 g **soaked seaweed**
- ☐ 10.5 oz / 300 g **beef brisket**
- ☐ 5 cups (34 oz / 1 L) **water**
- ☐ 1 tablespoon **soy sauce for soup**
- ☐ ½ tablespoon **fermented anchovy sauce**

- ☐ **건조 미역** 20g 혹은 **불린 미역** 150g
- ☐ **소고기 양지머리** 300g
- ☐ **물** 5컵 (1L)
- ☐ **국간장** 1큰술
- ☐ **멸치 액젓** ½큰술

SEASONING FOR BEEF

- ☐ 1½ tablespoons **soy sauce for soup**
- ☐ 1 teaspoon **toasted sesame oil**
- ☐ 1 teaspoon **minced garlic**

소고기 양념

- ☐ **국간장** 1½큰술
- ☐ **참기름** 1작은술
- ☐ **다진 마늘** 1작은술

Tips

- For beef, you can use the same amount of rump or round beef cuts instead of brisket. You can also replace beef with seafood such as dried pollack, oysters, clams, and abalones. However, clams should be left in water before use to remove sediment, and the total amount of seafood should be less than 7 oz / 200 g.
  how to remove sediment → page 180, #1
- You can use the same amount of fish sauce instead of soy sauce for soup and fermented anchovy sauce.

- 소고기는 양지머리 대신 같은 분량의 우둔살이나 홍두깨살을 사용해도 된다. 소고기 대신 황태, 굴, 조개, 전복 등 해산물을 사용해도 된다. 단, 조개는 해감한 뒤 사용해야 하며, 해산물의 전체 양은 200g 이내가 좋다.
  해감하는 방법 → 180 쪽, 1번
- 국간장과 멸치 액젓 대신 같은 분량의 피쉬 소스를 사용해도 된다.

Vegan 비건

- ☐ 4 tablespoons **perilla powder**
- ☐ 1 tablespoon **perilla oil**

- ☐ **들깻가루** 4큰술
- ☐ **들기름** 1큰술

Gluten-free 글루텐 프리

- ☐ 2½ tablespoons **gluten-free soy sauce for soup**
- ☐ ½ tablespoon **gluten-free fermented anchovy sauce**

- ☐ **글루텐 프리 국간장** 2½큰술
- ☐ **글루텐 프리 멸치 액젓** ½큰술

HOW TO COOK

1. Soak the dried seaweed in cold water for 15 minutes. Drain the seaweed through a strainer and cut it into bite-sized pieces.
2. Soak the beef in water for 5 minutes to remove the blood. Then cut it into bite-sized pieces.
3. Put the beef and all the seasoning ingredients for the beef in a bowl and mix well. Place the seasoned beef and prepared seaweed in a pot and stir-fry for 3 to 4 minutes over low heat.
   Vegan Leave out the beef. Add the prepared seaweed and 1 tablespoon of perilla oil to the pot and stir-fry for 3 to 4 minutes over low heat.
4. Pour water into the pot and boil for 30 minutes over medium heat. Season with soy sauce for soup and fermented anchovy sauce.
   Tip For a richer flavor, use the same amount of vegetable, anchovy, or beef broth instead of water. → **pages 219-221**
   Vegan Use vegetable broth instead of water and add perilla powder to the pot. Boil for 30 minutes. Leave out the fermented anchovy sauce and season with soy sauce for soup.
5. Simmer for 20 more minutes over low heat.
   Tip The umami taste seeps out from the seaweed when it is simmered slowly over low heat, so it tastes better than when it is boiled quickly over high heat.
   Tip If you refrigerate the remaining *miyeokguk* and boil it again later, the flavor will be richer.

1. 건조 미역은 찬물에 담가 15분간 불린다. 불려진 미역은 체에 밭쳐 물기를 제거하고, 먹기 좋은 크기로 자른다.
2. 소고기는 물에 담가 5분간 핏물을 제거하고, 먹기 좋은 크기로 썬다.
3. 볼에 소고기를 넣고, 소고기 양념 재료를 모두 넣어 버무린다. 양념한 고기와 손질한 미역을 냄비에 넣고 약불에서 3-4분간 볶는다.
   Vegan 소고기는 제외한다. 냄비에 손질한 미역과 들기름 1큰술을 넣고, 약불에서 3-4분간 볶는다.
4. 냄비에 물을 붓고 중불에서 30분간 끓인 뒤, 국간장과 멸치 액젓을 넣어 간을 맞춘다.
   Tip 물 대신 같은 분량의 채소 국물이나 멸칫국물, 고깃국물을 사용하면 맛이 더 풍부해진다. → **219-221** 쪽
   Vegan 냄비에 물 대신 채소 국물과 들깻가루를 넣어 30분간 끓인다. 멸치 액젓은 제외하고, 국간장으로 간을 맞춘다.
5. 약불에서 20분 이상 더 끓인다.
   Tip 약불에서 천천히 끓이면 미역의 감칠맛이 우러나 강불에서 빠르게 끓였을 때보다 더 맛있다.
   Tip 남은 미역국을 냉장 보관했다가 다시 한번 끓여 먹으면 처음 먹을 때보다 더 진한 맛을 느낄 수 있다.

Serves **2**
2인분

Difficulty **Mid**
난이도 **보통**

DISHES FOR SPECIAL FEASTS

BIRTHDAY

Japchae

Stir-Fried Glass Noodles and Vegetables

잡채

STORY ⟶ page **100**

INGREDIENTS

- ☐ 3.5 oz / 100 g dried glass noodles (*dangmyeon*) or 5 oz / 140 g soaked glass noodles
- ☐ 3.5 oz / 100 g beef rump
- ☐ ½ (3.5 oz / 100 g) onion
- ☐ ⅙ (2 oz / 60 g) carrot
- ☐ 1½ (1 oz / 30 g) oyster mushrooms
- ☐ 4 to 5 (0.7 oz / 20 g) scallions
- ☐ ⅓ teaspoon salt
- ☐ 2 tablespoons toasted sesame seeds
- ☐ 2 tablespoons cooking oil

- ☐ **건 당면** 100g 혹은 **불린 당면** 140g
- ☐ **소고기 우둔살** 100g
- ☐ **양파** ½개 (100g)
- ☐ **당근** ⅙개 (60g)
- ☐ **느타리버섯** 1½개 (30g)
- ☐ **쪽파** 4-5대 (20g)
- ☐ **소금** ⅓작은술
- ☐ **참깨** 2큰술
- ☐ **식용유** 2큰술

SEASONING FOR BEEF

- ☐ 1 tablespoon soy sauce
- ☐ 1 tablespoon toasted sesame oil
- ☐ 1 teaspoon sugar
- ☐ ½ teaspoon minced garlic
- ☐ ⅓ teaspoon black pepper

소고기 양념

- ☐ **간장** 1큰술
- ☐ **참기름** 1큰술
- ☐ **설탕** 1작은술
- ☐ **다진 마늘** ½작은술
- ☐ **후추** ⅓작은술

SEASONING FOR GLASS NOODLES

- ☐ 3 tablespoons soy sauce
- ☐ 2 tablespoons sugar
- ☐ 2 tablespoons toasted sesame oil
- ☐ ⅓ teaspoon black pepper

당면 양념

- ☐ **간장** 3큰술
- ☐ **설탕** 2큰술
- ☐ **참기름** 2큰술
- ☐ **후추** ⅓작은술

Vegan 비건

- ☐ 2 oz / 60 g various mushrooms (golden oak, button, wood ear mushrooms, etc.)
- ☐ **다른 종류의 버섯**(표고버섯, 양송이버섯, 목이버섯 등) 60g

Gluten-free 글루텐 프리

- ☐ 4 tablespoons gluten-free soy sauce
- ☐ **글루텐 프리 간장** 4큰술

Tips

- You can use the same amount of wide glass noodles or rice vermicelli (2.5 oz / 70 g) instead of glass noodles. When using rice vermicelli, put it in boiling water and boil for 7 minutes over medium heat. Then place it in a strainer before use.

- 당면 대신 같은 분량의 넓적한 당면이나 버미셀리 쌀국수 70g을 사용해도 된다. 버미셀리 쌀국수는 끓는 물에 넣고 중불에서 7분간 삶은 후 체에 밭쳐 사용한다.

- You can use the same amount of lean beef cuts such as round or bottom round instead of the rump. Also, you can use the same amount of seafood such as shrimp or squid instead of beef.
- You can use the same amount of golden oak, button, or wood ear mushrooms instead of oyster mushrooms.
- You can use the same amount of chives instead of scallions.

- 소고기는 우둔살 대신 같은 분량의 홍두깨살이나 설도살 등 지방이 적고 살코기가 많은 부위를 사용해도 된다. 혹은 고기 대신 같은 분량의 새우, 오징어 등 해산물을 사용해도 된다.
- 느타리버섯 대신 같은 분량의 표고버섯, 양송이버섯, 목이버섯 등을 사용해도 된다.
- 쪽파 대신 같은 분량의 부추를 사용해도 된다.

HOW TO COOK

1. Soak the dried glass noodles in cold water for 30 minutes, then drain them through a strainer.
2. Julienne the onion and carrot into 0.1-in/0.3-cm wide strips and cut the scallions into 2-in/5-cm lengths.
3. Remove the bottoms of the oyster mushrooms and tear them into strips by hand. Blanch the finely torn mushrooms in boiling water for 1 to 2 minutes.

   Vegan Cut the various mushrooms into bite-sized pieces and blanch them together with the oyster mushrooms for 1 to 2 minutes.
4. Thinly slice the beef into 0.2-in/0.5-cm wide pieces. Put all the seasoning ingredients for the beef into a bowl and mix well. Then add the beef and mix together.
5. Pour enough water to cook the noodles into a pot and bring it to a boil. When the water starts boiling, add the soaked glass noodles and boil for 3 to 4 more minutes. Take the noodles out using a strainer.
6. Add the cooking oil to a skillet and heat it. Stir-fry the julienned onion and carrot with the scallions, mushrooms, and salt for 1 to 2 minutes over high heat. Transfer them to a plate. Then add the seasoned beef to the skillet and stir-fry for 3 to 4 minutes over medium heat. Transfer it to a plate and set it aside.
7. Add the cooked glass noodles and the seasoning ingredients for glass noodles to the skillet and stir-fry for 2 to 3 minutes over low heat. But make sure to add the soy sauce and toasted sesame oil in small amounts at a time to taste.
8. Turn off the heat. Add all the blanched mushrooms, stir-fried vegetables, and beef, and mix well. Transfer the *japchae* to a plate and sprinkle toasted sesame seeds on top.

   Tip You can also make *japchaebap* (*japchae* with rice) by topping 7 oz / 200 g of cooked rice with *japchae* and seasoning it with 1 tablespoon of soy sauce and 1 tablespoon of toasted sesame oil.

1. 건 당면은 찬물에 담가 30분간 불리고, 체에 밭쳐 물기를 제거한다.
2. 양파와 당근은 0.3cm 폭으로 얇게 채 썰고, 쪽파는 5cm 길이로 썬다.
3. 느타리버섯은 밑동을 잘라내고 결을 따라 손으로 잘게 찢는다. 잘게 찢은 버섯은 끓는 물에 1-2분간 데친다.

   Vegan 여러 종류의 버섯을 모두 먹기 좋은 크기로 손질하여 느타리버섯과 함께 1-2분간 데친다.
4. 소고기는 0.5cm 폭으로 얇게 썬다. 볼에 소고기 양념 재료를 모두 넣어 고루 섞고, 소고기를 넣어 버무린다.
5. 면을 익히기 위해 냄비에 물을 적당량 붓고 끓인다. 물이 끓으면 불린 당면을 넣고 3-4분간 더 끓인 뒤, 당면을 체로 건진다.
6. 팬에 식용유를 두르고 불에 달군 뒤, 양파, 당근, 쪽파, 버섯과 소금을 넣고 강불에서 1-2분간 볶는다. 볶은 채소와 버섯을 접시에 옮겨 두고, 팬에 양념한 소고기를 넣어 중불에서 3-4분간 볶는다. 볶은 고기를 접시에 옮겨 둔다.
7. 팬에 익힌 당면과 당면 양념 재료를 넣은 뒤, 약불에서 2-3분간 볶는다. 이때, 간장과 참기름은 조금씩 넣어 가며 간을 맞춘다.
8. 불을 끄고 데친 버섯과 볶아 둔 채소, 고기를 모두 넣어 고루 섞는다. 접시에 옮겨 담고 참깨를 뿌린다.

   Tip 잡채를 밥 200g 위에 올리고, 간장 1큰술과 참기름 1큰술로 간을 맞추어 잡채밥을 만들어 먹어도 맛있다.

Japchae Recipe
Reference Video
잡채 레시피 참고 영상

Serves **2**
2인분

Difficulty **Low**
난이도 **쉬움**

- DISHES FOR SPECIAL FEASTS -

NEW YEAR'S DAY

Tteokguk

Sliced Rice Cake Soup

떡국

STORY ⟶ page **108**

INGREDIENTS

- ☐ 10.5 oz / 300 g sliced rice cakes
- ☐ 3.5 oz / 100 g beef brisket
- ☐ 3 eggs
- ☐ ¼ (4 in / 10 cm, 1 oz / 30 g) large green onion
- ☐ 0.1 oz / 2 g *gim* (dried seaweed)
- ☐ 1 tablespoon soy sauce for soup
- ☐ ½ teaspoon salt
- ☐ 5 cups (34 oz / 1 L) water
- ☐ 1½ tablespoons cooking oil

- ☐ **떡국떡** 300g
- ☐ **소고기 양지머리** 100g
- ☐ **계란** 3개
- ☐ **대파** ¼대 (10cm, 30g)
- ☐ **김** 2g
- ☐ **국간장** 1큰술
- ☐ **소금** ½작은술
- ☐ **물** 5컵 (1L)
- ☐ **식용유** 1½큰술

SEASONING FOR BEEF

- ☐ 1 tablespoon soy sauce
- ☐ 1 teaspoon minced garlic
- ☐ 1 teaspoon toasted sesame oil
- ☐ ½ teaspoon sugar
- ☐ ⅓ teaspoon black pepper

소고기 양념

- ☐ **간장** 1큰술
- ☐ **다진 마늘** 1작은술
- ☐ **참기름** 1작은술
- ☐ **설탕** ½작은술
- ☐ **후추** ⅓작은술

Tips

- You can use the same amount of lean beef cuts such as round or bottom round instead of brisket. Also, you can use the same amount of oysters, clams, or kimchi instead of beef.
- You can replace soy sauce for soup with the same amount of fermented anchovy sauce or fish sauce.

- 소고기는 양지머리 대신 같은 분량의 홍두깨살, 설도살 등 지방이 적고 살코기가 많은 부위를 사용해도 된다. 또는 소고기 대신 같은 분량의 굴, 조개, 김치 등을 넣어도 잘 어울린다.
- 국간장은 같은 분량의 멸치 액젓 또는 피쉬 소스로 대체할 수 있다.

Vegan 비건

- ☐ 2 (1.5 oz / 40 g) golden oak or button mushrooms
- ☐ **표고버섯 또는 양송이버섯** 2개 (40g)

Gluten-free 글루텐 프리

- ☐ 10.5 oz / 300 g gluten-free sliced rice cakes
- ☐ 1 tablespoon gluten-free soy sauce for soup
- ☐ 1 tablespoon gluten-free soy sauce (for seasoning for beef)
- ☐ **글루텐 프리 떡국떡** 300g
- ☐ **글루텐 프리 국간장** 1큰술
- ☐ **글루텐 프리 간장** 1큰술 (소고기 양념)

HOW TO COOK

1. Soak the sliced rice cakes in cold water for 15 minutes, and then drain them through a strainer.

   Tip To make sure the rice cakes are seasoned well, drain the soaked sliced rice cakes and mix them with ½ teaspoon of soy sauce.

2. Crack the eggs, separate the whites and yolks, and beat them separately. Add the cooking oil to a skillet and heat it. Then fry the white and yolks separately over medium heat into thin sheets. Cool the sheets and cut them into 0.1-in/0.2-cm wide and 1.5-in/4-cm long strips.

   Vegan Leave out the eggs.

3. Slice the large green onion diagonally into 0.1-in/0.2-cm thick pieces. Cut the *gim* into 0.2-in/0.5-cm thick strips with scissors.

4. Soak the beef in water for 5 minutes to remove the blood and cut it into bite-sized pieces. Mix all the seasoning ingredients for the beef in a bowl. Then add the beef and mix well. Place the seasoned beef in a pot and stir-fry for 3 minutes over low heat. Then pour in the water and bring it to a boil.

   Tip For a richer flavor, use the same amount of vegetable, anchovy, or beef broth instead of water. → **pages 219-221**

   Vegan Leave out the beef and prepare the vegetable broth.

5. After boiling the soup for 10 minutes over medium heat, add the soaked sliced rice cakes, large green onion, and soy sauce for soup, and boil for 15 more minutes. Then season with salt to taste.

   Vegan Thinly slice the golden oak or button mushrooms and boil them in the pot with the soaked sliced rice cakes and large green onion.

6. Transfer the complete *tteokguk* to a bowl and decorate it with the egg strips *and gim*.

   Tip Adding 4 to 5 steamed *mandu* (dumpling) on top of the *tteokguk* is also a good idea.

1. 떡국떡은 찬물에 담가 15분간 불리고, 체에 밭쳐 물기를 제거한다.

   Tip 불려서 물기를 제거한 떡국떡에 간장을 ½작은술 뿌려 버무려 두면 떡에 간이 잘 밴다.

2. 계란은 깨뜨려 흰자와 노른자를 분리한다. 흰자와 노른자를 각각 잘 푼다. 팬에 식용유를 두르고 달군 뒤, 중불에서 각각 지단을 부친다. 부친 지단은 한 김 식혀 0.2cm 폭에 4cm 길이로 썬다.

   Vegan 계란은 제외한다.

3. 대파는 0.2cm 두께로 어슷하게 썰고, 김은 가위를 사용하여 0.5cm 두께로 길쭉하게 자른다.

4. 소고기는 물에 담가 5분간 핏물을 제거하고, 먹기 좋은 크기로 썬다. 볼에 소고기 양념 재료를 모두 넣어 고루 섞고, 소고기를 넣어 버무린다. 양념한 고기를 냄비에 담고 약불에서 3분간 볶다가 물을 부어 끓인다.

   Tip 물 대신 같은 분량의 채소 국물이나 멸칫국물, 고깃국물을 사용하면 맛이 더 풍부해진다. → **219-221 쪽**

   Vegan 소고기는 제외하고, 채소 국물을 준비한다.

5. 중불에서 10분간 끓인 뒤, 불려 둔 떡국떡과 대파, 국간장을 넣어 15분간 더 끓인다. 소금을 넣어 간을 맞춘다.

   Vegan 표고버섯 또는 양송이버섯을 얇게 썰어 떡국떡, 대파와 함께 넣고 끓인다.

6. 완성된 떡국을 그릇에 옮겨 담고 잘라 둔 계란 지단과 김을 올려 장식한다.

   Tip 만두 4-5개를 찐 다음, 완성된 떡국에 얹어 먹어도 잘 어울린다.

Tteokguk Recipe
Reference Video
떡국 레시피 참고 영상

Serves **2 to 3**
2-3인분

Difficulty **Mid**
난이도 **보통**

DISHES FOR SPECIAL FEASTS

NEW YEAR'S DAY

Bulgogi

불고기

STORY ⟶ page **116**

INGREDIENTS

- ☐ 10.5 oz / 300 g thinly sliced beef sirloin
- ☐ 3 eggs
- ☐ 1 (7 oz / 200 g) onion
- ☐ ⅓ (4.3 oz / 120 g) carrot
- ☐ 2 (2.5 oz / 70 g) napa cabbage leaves
- ☐ ¼ (4 in / 10 cm, 1 oz / 30 g) large green onion
- ☐ 1½ (1 oz / 30 g) oyster mushrooms
- ☐ 1 (0.5 oz / 15 g) golden oak mushroom
- ☐ 2 to 3 (0.5 oz / 15 g) scallions

- ☐ 소고기 등심(불고기용) 300g
- ☐ 계란 3개
- ☐ 양파 1개 (200g)
- ☐ 당근 ⅓개 (120g)
- ☐ 배춧잎 2장 (70g)
- ☐ 대파 ¼대 (10cm, 30g)
- ☐ 느타리버섯 1½개 (30g)
- ☐ 표고버섯 1개 (15g)
- ☐ 쪽파 2-3대 (15g)

SEASONING FOR BEEF

- ☐ 2 (0.3 oz / 10 g) scallions
- ☐ 2 tablespoons soy sauce
- ☐ 1 tablespoon *cheongju* (refined rice wine)
- ☐ 1 tablespoon toasted sesame oil
- ☐ 1 teaspoon sugar
- ☐ ½ teaspoon minced garlic
- ☐ ⅓ teaspoon black pepper

소고기 양념

- ☐ 쪽파 2대 (10g)
- ☐ 간장 2큰술
- ☐ 청주 1큰술
- ☐ 참기름 1큰술
- ☐ 설탕 1작은술
- ☐ 다진 마늘 ½작은술
- ☐ 후추 ⅓작은술

ANCHOVY BROTH

3 cups (20.3 oz / 600 ml)

- ☐ 20 (0.7 oz / 20 g) dried anchovies for stock
- ☐ 1 sheet (2 in × 3 in / 5 cm × 7 cm) dried kelp
- ☐ 3½ cups (23.7 oz / 700 ml) water

멸칫국물

3컵 (600ml)

- ☐ 국물용 멸치 20마리 (20g)
- ☐ 건조 다시마 1장 (5cm × 7cm)
- ☐ 물 3½컵 (700ml)

Gluten-free 글루텐 프리

- ☐ 2 tablespoons gluten-free soy sauce
- ☐ 1 tablespoon gluten-free soy sauce for soup (for seasoning for broth)

- ☐ 글루텐 프리 간장 2큰술
- ☐ 글루텐 프리 국간장 1큰술 (국물 양념)

SEASONING FOR BROTH

- ☐ 1 tablespoon **soy sauce for soup**
- ☐ 1 teaspoon **salt**
- ☐ ½ teaspoon **minced garlic**
- ☐ ⅓ teaspoon **black pepper**

국물 양념

- ☐ **국간장** 1큰술
- ☐ **소금** 1작은술
- ☐ **다진 마늘** ½작은술
- ☐ **후추** ⅓작은술

Tips

- You can use the same amount of thinly sliced lean beef cuts such as chuck, tenderloin, and flank instead of sirloin.
- You can use the same amount of chives instead of scallions, and you can also add large green onions.
- You can use soju or dry white wine instead of *cheongju*.
- You can replace soy sauce for soup with the same amount of fermented anchovy sauce or fish sauce.
- Based on your preference, you can add boiled glass noodles (*dangmyeon*) or bite-sized rice cakes to the dish. You can also enjoy the broth with some rice.
- The dish tastes better if you can eat it as you simmer it over low heat at the table. This allows you to enjoy it warm.

- 소고기는 등심 대신 같은 분량의 목심, 안심, 설깃살 등 지방이 적고 살코기가 많은 부위를 얇게 썰어 사용해도 된다.
- 쪽파 대신 같은 분량의 부추를 사용하거나, 대파를 추가해도 된다.
- 청주 대신 소주 또는 달지 않은 화이트 와인을 사용해도 된다.
- 국간장은 같은 분량의 멸치 액젓 또는 피쉬 소스로 대체할 수 있다.
- 기호에 따라 삶은 당면이나 떡볶이 떡을 추가해도 잘 어울리고, 국물에 밥을 곁들여 먹어도 좋다.
- 식탁에서 약불에 끓여 가면서 먹으면 따뜻하게 즐길 수 있어서 더 맛있다.

HOW TO COOK

1. Slice the beef into bite-sized pieces. Pat the beef dry with a paper towel. Combine the beef and all the ingredients of the seasoning for the beef in a bowl and mix well. Place the bowl in the refrigerator to marinate for at least 30 minutes.
2. Prepare the anchovy broth. → **page 220**

   Tip You can use vegetable or beef broth instead of anchovy broth. → **pages 219-221** Also, you can use 3 cups of water instead.
3. Cut the onion into 0.2-in/0.5-cm thick strips, and julienne the carrot into 0.1-in/0.3-cm thick strips. Cut the napa cabbage leaves into 1.6 in × 2 in / 4 cm × 5 cm squares. Slice the large green onion diagonally into 0.1-in/0.3-cm thick pieces and finely chop the scallions.
4. Remove the bottoms of the oyster mushrooms and tear them into bite-sized pieces. Julienne the golden oak mushrooms into 0.1-in/0.3-cm thick strips.
5. Lay the cut napa cabbage leaves at the bottom of a pot and add the prepared vegetables and mushrooms on top. Pour in the anchovy broth and add all the seasoning ingredients for the broth. Let everything boil for 5 minutes over medium heat until the onion becomes translucent.

   Tip If you choose to use water instead of anchovy broth, use the same amount of fermented anchovy sauce or fish sauce instead of soy sauce for soup for the broth seasoning.
6. Adjust the heat to low. Spread the seasoned bulgogi into the pot and cook for 1 to 2 more minutes. Add the 3 eggs and put the lid on the pot. Then cook for 2 to 3 more minutes.

   Tip If you want to add boiled glass noodles or bite-sized rice cakes, put them in at this stage, before the beef, and cook them together. Since the glass noodles absorb the broth, it is best to add a moderate amount.
7. Turn off the heat. Sprinkle the chopped scallions evenly into the pot.

1. 소고기는 먹기 좋게 썬 다음, 키친타월을 사용하여 핏물을 제거한다. 볼에 소고기 양념 재료를 모두 넣고, 소고기를 넣어 버무린다. 양념한 고기는 냉장고에 넣어 30분 이상 숙성시킨다.
2. 멸칫국물을 준비한다. → **220쪽**

   Tip 멸칫국물 대신 채소 국물이나 고깃국물을 사용해도 좋다. → **219-221쪽** 또는 물 3컵을 사용해도 된다.
3. 양파는 0.5cm 두께로 채 썰고, 당근은 0.3cm 두께로 얇게 채 썬다. 배춧잎은 4cm × 5cm 크기로 네모나게 썬다. 대파는 0.3cm 두께로 어슷하게 썰고, 쪽파는 잘게 다진다.
4. 느타리버섯은 밑동을 잘라 내어 먹기 좋은 크기로 찢고, 표고버섯은 0.3cm 두께로 얇게 채 썬다.
5. 냄비에 잘라 둔 배춧잎을 깔고 손질한 채소와 버섯을 올린 뒤, 멸칫국물을 붓고 국물 양념 재료를 모두 넣는다. 중불에서 양파가 투명해질 때까지 5분간 끓인다.

   Tip 멸칫국물 대신 물을 사용한다면, 국물 양념의 국간장 대신 같은 분량의 멸치 액젓이나 피쉬 소스를 사용한다.
6. 약불로 줄인 뒤, 양념한 불고기를 펼쳐 넣고 1-2분간 더 익힌다. 계란 3개를 깨뜨려 넣은 다음, 뚜껑을 덮고 2-3분간 더 익힌다.

   Tip 삶은 당면이나 떡볶이 떡을 추가한다면 이 단계에서 고기보다 먼저 넣어 고기와 함께 익힌다. 당면은 국물을 흡수하므로 적당량만 넣는 것이 좋다.
7. 불을 끄고 다진 쪽파를 고루 뿌린다.

MEMO

6

Perfect Pairings with Drinks

술과 어울리는 안주

Kimchijeon 김치전
Gyeran Mari 계란말이

Makes **2 pancakes** (8.7-in/22-cm in diameter)
2장 (지름 22cm)

Difficulty **Low**
난이도 **쉬움**

- PERFECT PAIRINGS WITH DRINKS -

Kimchijeon

Kimchi Pancake

김치전

STORY ⟶ page **126**

INGREDIENTS

- ☐ 1¼ cups (10.5 oz / 300 g) **kimchi** (weight excluding the kimchi juice)
- ☐ 1¼ cups (5.7 oz / 160 g) **all-purpose flour**
- ☐ ¼ (1.7 oz / 50 g) **onion**
- ☐ 1 cup (7 oz / 200 ml) **water**
- ☐ 2 tablespoons **kimchi juice**
- ☐ 1 teaspoon **salt**
- ☐ 4 tablespoons **cooking oil**

- ☐ **김치** 1¼컵 (300g) (김칫국물 제외한 무게)
- ☐ **밀가루(중력분)** 1¼컵 (160g)
- ☐ **양파** ¼개 (50g)
- ☐ **물** 1컵 (200ml)
- ☐ **김칫국물** 2큰술
- ☐ **소금** 1작은술
- ☐ **식용유** 4큰술

Vegan 비건

- ☐ 1¼ cups (10.5 oz / 300 g) **vegan kimchi** (weight excluding the kimchi juice)
- ☐ 2 tablespoons **vegan kimchi juice**

- ☐ **비건 김치** 1¼컵 (300g) (김칫국물 제외한 무게)
- ☐ **비건 김칫국물** 2큰술

Gluten-free 글루텐 프리

- ☐ 1¼ cups (10.5 oz / 300 g) **gluten-free kimchi** (weight excluding the kimchi juice)
- ☐ 1¼ cups (5.7 oz / 160 g) **gluten-free wheat flour**
- ☐ 2 tablespoons **gluten-free kimchi juice**

- ☐ **글루텐 프리 김치** 1¼컵 (300g) (김칫국물 제외한 무게)
- ☐ **글루텐 프리 밀가루** 1¼컵 (160g)
- ☐ **글루텐 프리 김칫국물** 2큰술

Tips

- You can use 1.5 oz / 40 g of scallions or chives instead of onion.
- If you use Korean pancake mix instead of all-purpose flour, for 1¼ cups (5.7 oz / 160 g) of the mix, use only 9⁄10 cup (6 oz / 180 ml) of water and leave out the salt.
- Based on your preferences, you can add scallions, pork belly (*samgyeopsal*), squid, shrimp, or mozzarella cheese to the batter.

- 양파 대신 쪽파 혹은 부추 40g을 사용해도 된다.
- 밀가루(중력분) 대신 부침가루를 사용한다면 부침가루 1¼컵 (160g) 기준으로 물은 9⁄10컵 (180ml)만 사용하고, 소금은 제외한다.
- 기호에 따라 반죽에 쪽파, 돼지고기 삼겹살, 오징어, 새우, 모차렐라 치즈를 추가해도 잘 어울린다.

HOW TO COOK

1. Drain the kimchi and chop it finely. Julienne the onion into 0.1-in/ 0.2-cm thick strips.
2. Add the flour and water to a bowl and mix well. Then add the kimchi and onion, and mix again. Once everything is mixed well, add the kimchi juice and salt, and mix again.
3. Add 2 tablespoons of cooking oil to a skillet and heat it. Spread a big ladle of batter evenly onto the skillet.
   Tip If the batter is too watery, add flour. If the batter is too thick, add water. You may also add kimchi juice instead of water.
4. Cook for 3 minutes over medium heat, then flip it over and cook for another 2 minutes.

1. 김치는 물기를 짜내고 잘게 썬다. 양파는 0.2cm 두께로 얇게 채 썬다.
2. 볼에 밀가루와 물을 넣고 고루 푼 다음, 썰어 둔 김치와 양파를 넣고 섞는다. 반죽이 잘 섞인 뒤, 김칫국물과 소금을 넣고 고루 섞는다.
3. 팬에 식용유 2큰술을 두르고 달군 뒤, 반죽을 한 국자 크게 떠서 고르게 펼친다.
   Tip 반죽이 묽게 되었다면 밀가루를, 뻑뻑하게 되었다면 물을 추가한다. 물 대신 김칫국물을 추가해도 좋다.
4. 중불에서 3분간 구운 뒤, 뒤집어서 2분간 더 굽는다.

Serves **2**
2인분

Difficulty **Low**
난이도 **쉬움**

- PERFECT PAIRINGS WITH DRINKS -

Gyeran Mari

Rolled Omelet

계란말이

STORY ⟶ page **134**

INGREDIENTS

- ☐ 4 eggs
- ☐ ⅙ (2 oz / 60 g) carrot
- ☐ ⅙ (1 oz / 30 g) onion
- ☐ 2 (0.3 oz / 10 g) scallions
- ☐ 1 slice cheese
- ☐ ½ teaspoon salt
- ☐ a dash of ketchup
- ☐ 3 tablespoons cooking oil

- ☐ **계란** 4개
- ☐ **당근** ⅙개 (60g)
- ☐ **양파** ⅙개 (30g)
- ☐ **쪽파** 2대 (10g)
- ☐ **슬라이스 치즈** 1장
- ☐ **소금** ½작은술
- ☐ **케첩** 약간
- ☐ **식용유** 3큰술

Vegan 비건

- ☐ 7 oz / 200 g vegan eggs
- ☐ 1 slice vegan cheese

- ☐ **비건 계란** 200g
- ☐ **비건 슬라이스 치즈** 1장

Gluten-free 글루텐 프리

- ☐ a dash of gluten-free ketchup

- ☐ **글루텐 프리 케첩** 약간

Special Tip <Run BTS!> EP. 102

You can use the same amount of chives instead of scallions.
쪽파 대신 같은 분량의 부추를 사용해도 돼요.

Tips

- The dish is still delicious if you make it with just eggs, without the vegetables or cheese.
- Adding 1 tablespoon of *cheongju* (refined rice wine) to the egg mixture will make it taste smoother.
- Adding 0.3-0.7 oz / 10-20 g of finely chopped broccoli, perilla leaves (*kkaennip*), ham, and mushrooms to the egg mixture is also a good idea. However, be careful not to add too many ingredients in proportion to the egg mixture, as it will be hard to roll up neatly and can lead to cracks.

- 채소나 치즈를 넣지 않고 계란만 사용하여 만들어도 맛있다.
- 계란물에 청주를 1큰술 추가하면 더 부드러운 맛을 낼 수 있다.
- 브로콜리, 깻잎, 햄, 버섯을 10-20g 정도 다져 넣어도 잘 어울린다. 단, 계란물의 양에 비해 재료가 너무 많이 들어가면 예쁘게 말아지지 않고 갈라질 수 있으니 주의한다.

HOW TO COOK

1. Chop the carrot, onion, and scallions very finely.
2. Crack the eggs into a bowl, beat them well, and add the chopped vegetables and salt.
3. Add the cooking oil to a skillet and heat it. Pour in the egg mixture and spread it out into a thin layer over medium heat. Cut a slice of cheese into 2-in/5-cm wide pieces and arrange them in a single row. Adjust the heat to low. Once the edges start to cook, starting from one end, fold it into a 2 to 3-in / 5 to 8-cm wide roll.
   photo → page 138
   Tip When pouring the egg mixture into the skillet for the first time, it is best to spread it into a thin layer, as a thick layer is hard to fold.
4. Push the rolled egg to one side of the skillet and pour in the remaining egg mixture on the other side to make one big roll. Once the edges are cooked, keep rolling in one direction and cook for 2 to 3 more minutes over low heat.
5. Transfer the *gyeran mari* to a cutting board, and let it cool. Then cut it into bite-sized pieces. Enjoy it with ketchup.

1. 당근, 양파, 쪽파를 아주 잘게 다진다.
2. 볼에 계란을 깨뜨려 넣고 잘 푼 다음, 다진 채소와 소금을 넣어 섞는다.
3. 팬에 식용유를 두르고 달군 뒤, 중불에서 계란물을 부어 얇게 펼친다. 슬라이스 치즈를 5cm 폭으로 잘라 한 줄로 길게 올린다.
   약불로 줄인 다음, 가장자리가 익어 가면 한쪽 가장자리에서부터 5-8cm 정도 폭으로 접듯이 만다. 사진 → **138 쪽**
   Tip 계란이 너무 두꺼우면 잘 접히지 않으므로, 처음 팬에 올릴 때는 얇게 펼치는 것이 좋다.
4. 말아 둔 계란을 팬 한쪽으로 밀고, 다른 쪽에 남은 계란물을 마저 붓는다. 가장자리가 익어 가면 한 방향으로 굴려 가며 마저 말고, 약불에서 2-3분간 더 익힌다.
5. 도마에 올려 잠시 식힌 뒤 먹기 좋은 크기로 썬다. 먹을 때 케첩을 곁들인다.

Special Tip <Run BTS!> EP. 102

Use a spatula with a spoon to roll the *gyeran mari* more easily.
계란말이를 말 때에는 뒤집개와 숟가락을 함께 사용하면 더 쉽게 말 수 있어요.

7

Sweet Desserts

달달한 디저트

Dalgona 달고나

Hwachae 화채

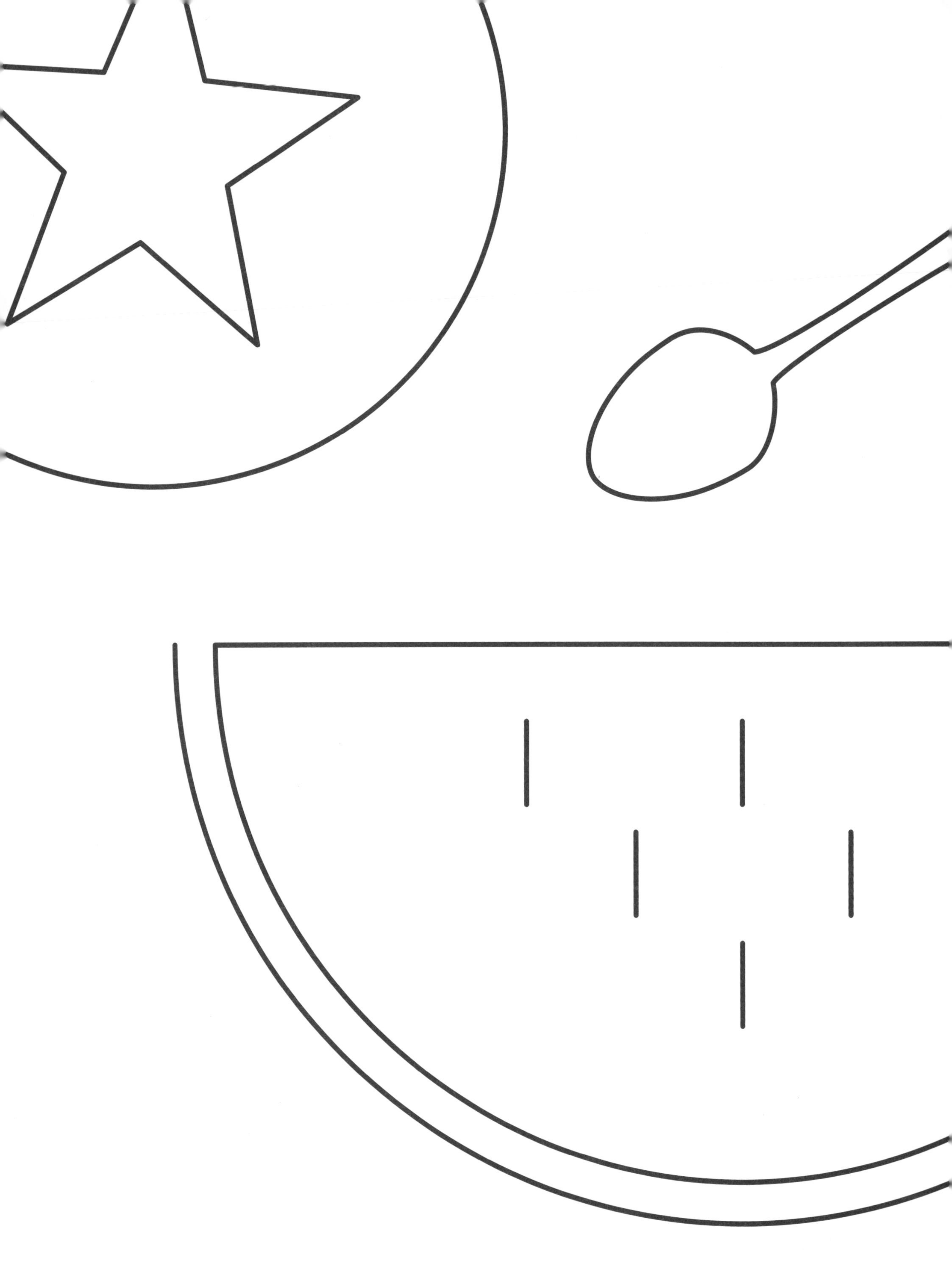

Makes **1 piece**
(4-in/10-cm in diameter)
1개 (지름 10cm)

Difficulty **Low**
난이도 **쉬움**

Dalgona

Sugar Candy

달고나

STORY ⟶ page **144**

INGREDIENTS

- ☐ 2 tablespoons **sugar**
- ☐ ½ teaspoon **baking soda**
- ☐ a dash of **cooking oil**

- ☐ **설탕** 2큰술
- ☐ **식용 베이킹 소다** ½작은술
- ☐ **식용유** 약간

TOOLS

- ☐ tray
- ☐ 1 wooden chopstick
- ☐ ladle
- ☐ cookie cutters
- ☐ press

도구

- ☐ **쟁반**
- ☐ **나무젓가락** 1개
- ☐ **국자**
- ☐ **모양틀**
- ☐ **누르개**

Tips

- Use a stainless steel tray. Chopsticks that are made of wood are the safest to use.
- 쟁반은 스테인리스 재질로 된 것을 사용한다. 젓가락은 나무젓가락을 사용하는 것이 안전하다.

* This dish is vegan and gluten-free.

* 비건, 글루텐 프리 요리

HOW TO COOK

1. Pour the sugar into a ladle and melt it over the lowest heat as you stir it with a wooden chopstick.
   Tip Adjust the distance between the heat and the ladle so that neither the sugar nor the ladle burns.
2. When the sugar melts and becomes transparent, dip the chopstick in the baking soda a couple of times, add it to the sugar, and stir quickly.
   Tip Add the appropriate amount of baking soda. Too little baking soda will make the melted sugar too sticky, which will make it difficult to separate from the press. Too much baking soda will cause it to puff up and may add a bitter taste.
3. Pour the *dalgona* onto a flat tray. Push the press onto the *dalgona*, gently flatten it, and lift up the press. Then set the cookie cutter on top of the *dalgona* and push it with the press to stamp a shape.
   photo → page 147
   Tip Before pouring the *dalgona* onto the tray, put some cooking oil on a paper towel and use it to apply a coat of oil on the tray, press, and cookie cutter. This will help the *dalgona* slide off easily.
4. Let it cool for a while. Then enjoy the *dalgona* starting from the edges as you try to pop the shape out.
   Tip Before you do the dishes after making *dalgona*, make sure to soak all the tools you used in water until the *dalgona* completely melts away.

   Tip How to make a *dalgona* latte
   Make a thick layer of *dalgona* and crumble it. Place the crumbs on top of an iced café latte. Take a mouthful of the *dalgona* with your drink to enjoy the sweet and crunchy *dalgona* latte.

1. 국자에 설탕을 담고, 가장 약한 불에서 나무젓가락으로 저어 가며 녹인다.
   Tip 설탕이나 국자가 타지 않도록 불과 국자 사이 거리를 조절한다.
2. 설탕이 녹아 투명해지면 베이킹 소다를 젓가락으로 두어 번 콕콕 찍어 넣고 빠르게 젓는다.
   Tip 베이킹 소다는 적당량만 넣는다. 베이킹 소다가 너무 적게 들어가면 끈적거려서 누르개에서 잘 떨어지지 않고, 너무 많이 들어가면 부풀어 오르고 쓴맛이 날 수 있다.
3. 평평한 쟁반에 완성된 달고나를 붓고 누르개로 살짝 눌렀다 떼어 납작하게 만든다. 모양틀을 얹은 다음, 그 위를 누르개로 한 번 더 눌러 모양을 찍어 낸다. **사진 → 147 쪽**
   Tip 쟁반에 달고나를 붓기 전에, 키친타월에 식용유를 묻혀 쟁반과 누르개, 모양틀을 기름으로 한 차례 코팅해 두면 달고나가 달라붙지 않고 잘 떨어진다.
4. 완성된 달고나는 한 김 식힌 뒤, 모양을 따라 떼어 내며 먹는다.
   Tip 달고나를 만들고 나서 설거지를 할 때는 사용한 도구를 물에 충분히 담가 두었다가 달고나가 모두 녹아 사라진 다음 씻는다.

   Tip **달고나 라떼 만들기**
   달고나를 두툼하게 만들어 잘게 부순 다음 시원한 카페 라떼 1잔에 얹는다. 달고나와 음료를 한입에 마시면 달콤하고 바삭한 맛의 달고나 라떼를 즐길 수 있다.

Dalgona Recipe
Reference Video
달고나 레시피 참고 영상

Special Tip <Run BTS!> EP. 95

It's fun to eat the *dalgona* stuck on the ladle or chopstick after you've poured the *dalgona* on the tray.
달고나를 쟁반에 붓고 나서 국자나 젓가락에 붙은 달고나를 떼어 먹는 것도 재미있어요.

Serves **2**
2인분

SWEET DESSERTS

Difficulty **Low**
난이도 **쉬움**

Hwachae

Punch

화채

STORY ⟶ page **152**

INGREDIENTS

- ☐ ⅓ (2 lb / 1 kg) **watermelon**
- ☐ 3.5 oz / 100 g **canned peaches**
- ☐ 1 can (12 oz / 355 ml) **soda**
- ☐ 1½ cups (10 oz / 300 ml) **milk**
- ☐ ½ cup (3.5 oz / 100 ml) **peach juice from the can**

- ☐ **수박** ⅓통 (1kg)
- ☐ **복숭아 통조림** 100g
- ☐ **사이다** 1캔 (355ml)
- ☐ **우유** 1½컵 (300ml)
- ☐ **복숭아 통조림 물** ½컵 (100ml)

Vegan 비건

- ☐ 1½ cups (10 oz / 300 ml) **vegan milk**
- ☐ **비건 우유** 1½컵 (300ml)

* This dish is gluten-free.
* 글루텐 프리 요리

HOW TO COOK

1. Scoop out 2 lb / 1 kg of the insides of the watermelon with a spoon or a small ice cream scoop, and put it into a bowl.
   Tip You can cut the watermelon into bite-sized pieces instead of scooping it out.
2. Open the can of peaches and set aside ½ cup (3.5 oz / 100 ml) of peach juice from the can. Take the peaches out and cut them into bite-sized pieces.
3. Put the watermelon, peaches, soda, milk, and peach juice into the bowl and mix well.
   Tip Add the soda and milk a little at a time to taste.

1. 수박 속을 숟가락 또는 작은 사이즈의 아이스크림 스쿱으로 1kg만큼 파내고 볼에 담는다.
   Tip 수박은 파내지 않고 한입 크기로 썰어서 준비해도 된다.
2. 복숭아 통조림을 열고, 통조림 물을 ½컵 (100ml) 덜어 둔다. 복숭아는 건져내어 먹기 좋은 크기로 썬다.
3. 볼에 수박, 복숭아, 사이다, 우유, 복숭아 통조림 물을 넣어 잘 섞는다.
   Tip 사이다와 우유는 조금씩 넣어 가며 기호에 따라 양을 조절한다.

Tips

- If the watermelon is not sweet enough, add sugar to taste.
- Aside from canned peaches, you can add other seasonal fruits, such as strawberries, pineapples, kiwis, and lychees.
- Add ice cubes to make the drink even cooler.

- 수박의 당도가 낮으면 설탕을 추가해도 좋다.
- 복숭아 통조림 외에 딸기, 파인애플, 키위, 리치 등 다른 계절 과일을 넣어도 잘 어울린다.
- 얼음을 추가하면 더 시원하게 즐길 수 있다.

Hwachae Recipe
Reference Video
화채 레시피 참고 영상

MEMO

Basic Preparations

기본 조리법

Rice 밥

Vegetable Broth 채소 국물

Anchovy Broth 멸칫국물

Beef Broth 고깃국물

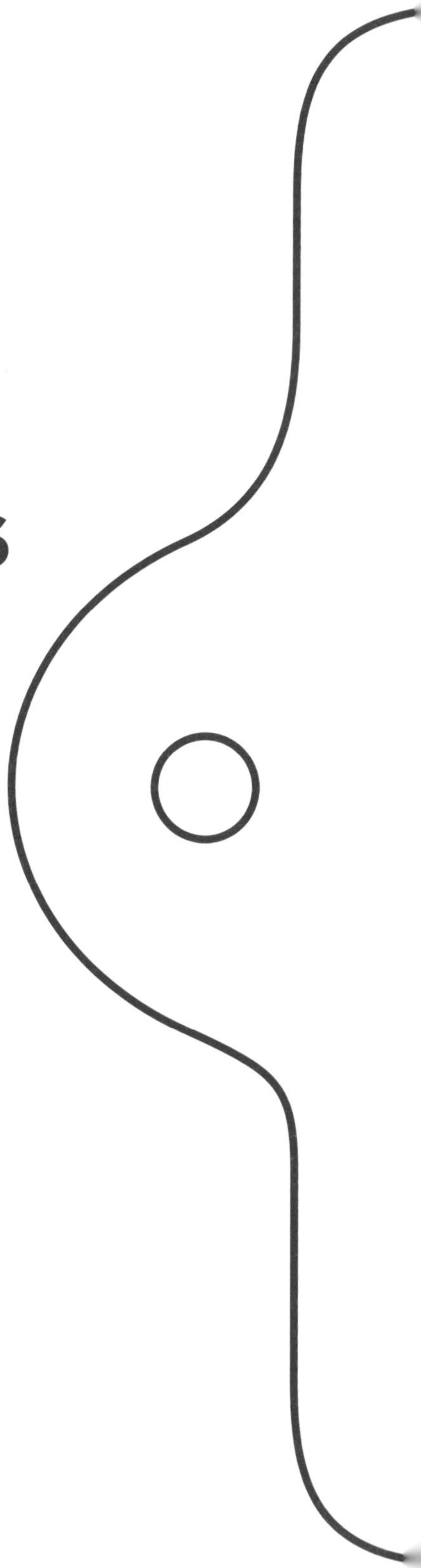

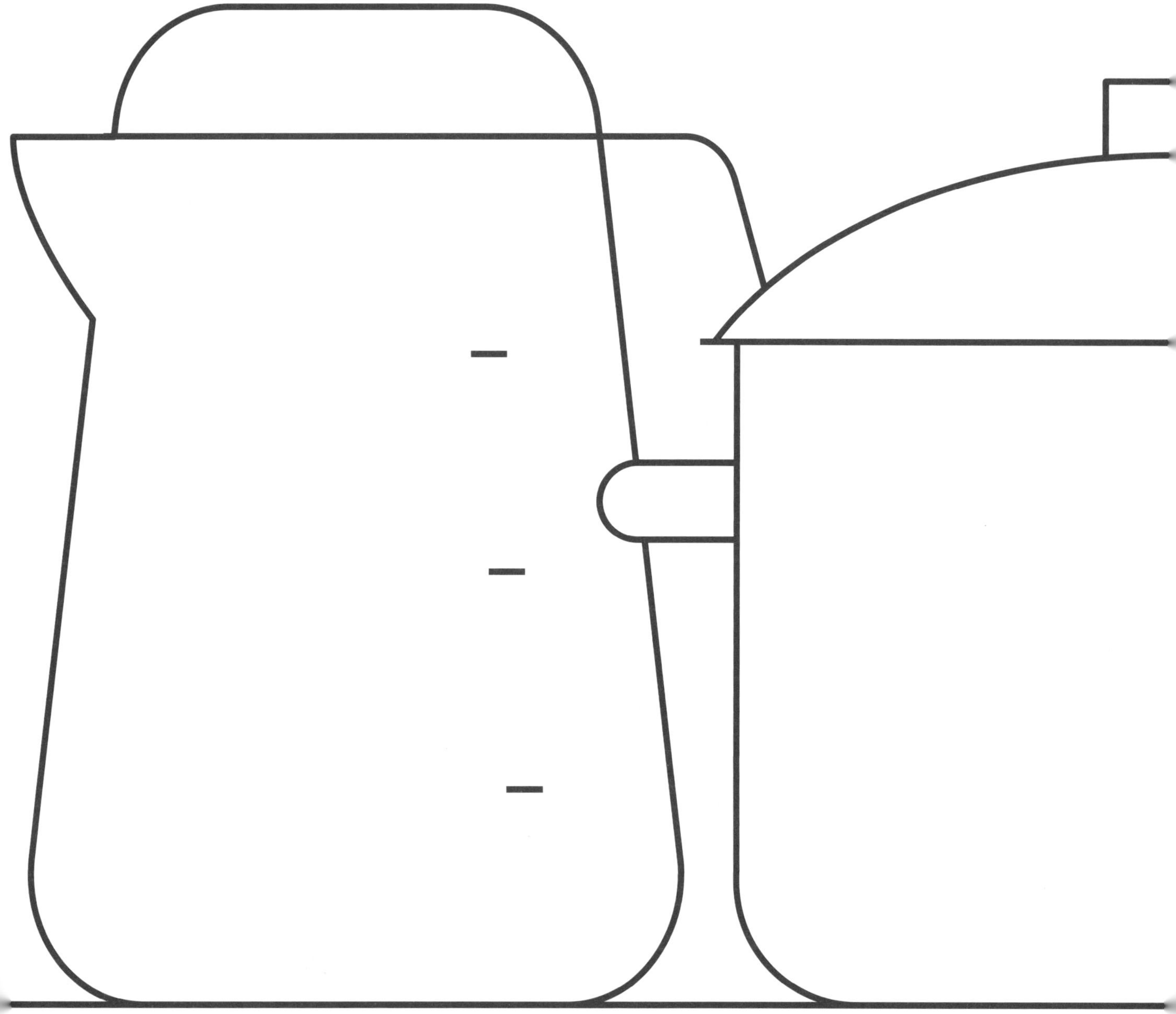

Serves **2**
2인분

Difficulty **Low**
난이도 **쉬움**

- BASIC PREPARATIONS -

Rice

밥

INGREDIENTS

☐ 1⅕ cups (7.7 oz / 220 g) **uncooked rice**
☐ 1½ cups (10 oz / 300 ml) **water**

☐ **쌀** 1⅕컵 (220g)
☐ **물** 1½컵 (300ml)

HOW TO COOK

1. Put the rice in a bowl and pour in just enough water to cover it.
2. Swoosh the rice around to wash it. Rinse it 2 to 3 times and drain it with a strainer.
3. Put the rinsed rice and 1½ cups of water in a pot and bring to a boil over high heat, stirring with a wooden spoon.
   Tip Using a pot with a thick bottom is best for cooking rice.
4. When the water starts boiling, close the lid on the pot and simmer over the lowest heat for 12 minutes.
   Tip Because it is most important to let the rice cook fully in the residual heat, avoid opening the lid between steps 4 and 5.
5. Turn the heat off and let it sit for 10 minutes with the lid still closed.
6. Open the lid to check if the rice is done and fluff the rice.

1. 볼에 쌀을 넣고, 쌀이 잠길 정도의 물을 붓는다.
2. 쌀을 주무르듯이 씻어 2-3번 헹구고, 체에 밭쳐 물기를 뺀다.
3. 냄비에 씻은 쌀과 물 1½컵을 넣고, 강불에서 주걱으로 저어 가며 끓인다.
   Tip 밥 짓는 냄비는 가능하면 바닥이 두꺼운 것을 사용한다.
4. 물이 끓으면 뚜껑을 덮고, 제일 약한 불에서 12분간 익힌다.
   Tip 밥을 지을 때는 냄비 속 열기로 쌀을 완전히 익히는 것이 가장 중요하므로 4에서 5단계로 넘어갈 때 뚜껑을 열어 보지 않도록 주의한다.
5. 불을 끄고, 뚜껑을 닫은 채로 10분간 뜸을 들인다.
6. 뚜껑을 열어 밥이 다 되었는지 확인하고, 밥을 가볍게 섞는다.

* This dish is vegan and gluten-free.
* 비건, 글루텐 프리 요리

Tips

- Any type of rice can be used. Most Koreans use sticky rice, but nonsticky jasmine or basmati rice are also good options. However, make sure to use 1⅔ cups (11 oz / 330 ml) of water for either of these alternative rices.
- 쌀은 어느 종류를 사용해도 괜찮다. 한국에서는 주로 찰기 있는 쌀을 사용하지만, 찰기가 없는 재스민 쌀이나 바스마티 쌀을 사용해도 좋다. 단, 재스민 쌀이나 바스마티 쌀을 사용한다면 물의 양을 1⅔컵(330ml)으로 늘린다.

5 cups (34 oz / 1 L)
5컵 (1L)

BASIC PREPARATIONS

Difficulty **Low**
난이도 **쉬움**

Vegetable Broth

채소 국물

INGREDIENTS

- ☐ 1 (4.3 oz / 120 g) **large green onion** (white part and root)
- ☐ ½ (3.5 oz / 100 g) **onion**
- ☐ 1 piece (0.8 in / 2 cm, 3 oz / 100 g) **Korean radish**
- ☐ 10 (1 oz / 30 g) **dried golden oak mushrooms**
- ☐ 1 sheet (2 in × 3 in / 5 cm × 7 cm) **dried kelp**
- ☐ 7½ cups (50.7 oz / 1.5 L) **water**

- ☐ **대파** 1대 (120g) (흰 부분과 뿌리)
- ☐ **양파** ½개 (100g)
- ☐ **무** 1조각 (2cm, 100g)
- ☐ **건조 표고버섯** 10개 (30g)
- ☐ **건조 다시마** 1장 (5cm × 7cm)
- ☐ **물** 7½컵 (1.5L)

HOW TO COOK

1. Wash the dried golden oak mushrooms in running water. Place the washed mushrooms and 7½ cups (50.7 oz / 1.5 L) of water in a bowl and let the mushrooms soak for at least 30 minutes until they become soft.
2. Transfer the mushrooms and mushroom-soaked water to a pot, add the large green onion, onion, Korean radish, and dried kelp, and bring to a boil. When the water starts to boil, take out the kelp and let it simmer for at least 20 minutes. Strain the finished broth before use.
   Tip Once the water starts to boil, take out the kelp right away, as boiling it too long will release a slimy substance.

1. 흐르는 물에 건조 표고버섯을 씻는다. 씻은 버섯과 물 7½컵 (1.5L)을 볼에 넣고 표고버섯이 부드러워질 때까지 30분 이상 불린다.
2. 버섯과 버섯 불린 물을 냄비로 옮겨 담은 뒤, 대파, 양파, 무와 건조 다시마를 넣고 끓인다. 물이 끓으면 다시마를 먼저 건지고, 20분 이상 더 끓인다. 완성된 국물은 체에 걸러 사용한다.
   Tip 다시마는 오래 끓이면 끈적이는 성분이 나오기 때문에 물이 팔팔 끓으면 바로 건져 낸다.

* This dish is vegan and gluten-free.

* 비건, 글루텐 프리 요리

Tips

- If the Korean radish is unavailable, you can leave it out.
- Golden oak mushrooms are the best, but if they are unavailable, you can use other mushrooms, too.

- 무를 구하기 어렵다면 제외해도 된다.
- 버섯은 표고버섯이 가장 좋으나, 표고버섯을 구하기 어렵다면 다른 버섯을 사용해도 된다.

5 cups (34 oz / 1 L)
5컵 (1L)

- BASIC PREPARATIONS -

Difficulty **Low**
난이도 **쉬움**

Anchovy Broth

멸칫국물

INGREDIENTS

- ☐ 40 (1.5 oz / 40g) dried anchovies for stock
- ☐ 1 sheet (2 in × 3 in / 5 cm × 7 cm) dried kelp
- ☐ 6 cups (40.5 oz / 1.2 L) water

- ☐ **국물용 멸치** 40마리 (40g)
- ☐ **건조 다시마** 1장 (5cm × 7cm)
- ☐ 물 6컵 (1.2L)

HOW TO COOK

1. Remove the heads and intestines of the dried anchovies and cook them in a pot over low heat for 2 to 3 minutes.
2. Add the dried kelp and 6 cups (40.5 oz / 1.2 L) of water to the pot and bring to a boil. When the water starts to boil, take out the kelp and simmer for 5 minutes. Then take out the anchovies. Strain the finished broth before use.
   Tip Once the water starts to boil, take out the kelp right away, as boiling it too long will release a slimy substance.

1. 국물용 멸치의 머리와 내장을 제거하고 냄비에 담아 약불에서 2-3분간 볶는다.
2. 냄비에 건조 다시마와 물 6컵 (1.2L)을 넣고 끓인다. 물이 끓으면 다시마를 건지고, 5분 뒤에 멸치를 건진다. 완성된 국물은 체에 걸러 사용한다.
   Tip 다시마는 오래 끓이면 끈적이는 성분이 나오기 때문에 물이 팔팔 끓으면 바로 건져 낸다.

* This dish is gluten-free.
* 글루텐 프리 요리

Tips

- The white powdery substance on the dried kelp is the main source of umami, so use it as is without washing it off.
- 건조 다시마에 묻은 흰 가루는 감칠맛의 주성분이므로 씻지 않고 그대로 넣는다.

5 cups (34 oz / 1 L)
5컵 (1L)

- BASIC PREPARATIONS -

Difficulty **Low**
난이도 **쉬움**

Beef Broth

고깃국물

INGREDIENTS

- ☐ 10.5 oz / 300 g **beef brisket**
- ☐ 1 (4.3 oz / 120 g) **large green onion** (white part and root)
- ☐ 1 piece (0.8 in / 2 cm, 3 oz / 100 g) **Korean radish**
- ☐ 3 cloves **garlic**
- ☐ 7½ cups (50.7 oz / 1.5 L) **water**

- ☐ **소고기 양지머리** 300g
- ☐ **대파** 1대 (120g) (흰 부분과 뿌리)
- ☐ **무** 1조각 (2cm, 100g)
- ☐ **마늘** 3톨
- ☐ **물** 7½컵 (1.5L)

HOW TO COOK

1. Soak the beef in water for 10 minutes to remove the blood.
2. Put the beef, 7½ cups (50.7 oz / 1.5 L) of water, and vegetables into a pot and simmer for 50 minutes over medium heat. Strain the finished broth before use.

1. 소고기는 물에 10분간 담가 핏물을 제거한다.
2. 냄비에 소고기와 물 7½컵 (1.5L), 채소를 모두 넣고 중불에서 50분간 끓인다. 완성된 국물은 체에 걸러 사용한다.

* This dish is gluten-free.

* 글루텐 프리 요리

Tips

- You can use the same amount of lean beef cuts such as round or bottom round instead of brisket.
- If the Korean radish is unavailable, you can leave it out.

- 소고기는 양지머리 대신 지방이 적고 살코기가 많은 홍두깨살, 설도살 등을 사용해도 된다.
- 무를 구하기 어렵다면 제외해도 된다.

Ingredients Index

	Ingredients		Page	Dishes	
A	abalones	전복	194	Miyeokguk	미역국
	anchovies (dried anchovies for stock)	국물용 멸치	220	Anchovy Broth	멸칫국물
B	bacon	베이컨	174	Toast (j-hope's)	제이홉표 토스트
	baking soda	식용 베이킹 소다	212	Dalgona	달고나
	beef boneless beef ribs	소고기 갈빗살	182	Jjajang Ramyeon	짜장 라면
	beef bottom round	소고기 설도살	196	Japchae	잡채
			198	Tteokguk	떡국
			221	Beef Broth	고깃국물
	beef brisket	소고기 양지머리	194	Miyeokguk	미역국
			198	Tteokguk	떡국
			221	Beef Broth	고깃국물
	beef chuck	소고기 목심	200	Bulgogi	불고기
	beef flank	소고기 설깃살	200	Bulgogi	불고기
	beef round beef cuts	소고기 홍두깨살	194	Miyeokguk	미역국
			196	Japchae	잡채
			198	Tteokguk	떡국
			221	Beef Broth	고깃국물
	beef rump	소고기 우둔살	194	Miyeokguk	미역국
			196	Japchae	잡채
	beef sirloin	소고기 등심	164	Ganjang Gyeran Bap	간장계란밥
			182	Jjajang Ramyeon	짜장 라면
	beef sirloin (thinly sliced)	소고기 등심(불고기용)	200	Bulgogi	불고기
	beef tenderloin	소고기 안심	182	Jjajang Ramyeon	짜장 라면
			200	Bulgogi	불고기
	bread (sliced)	식빵	174	Toast (j-hope's)	제이홉표 토스트
			175	Toast (Jung Kook's)	정국표 토스트
	broccoli	브로콜리	208	Gyeran Mari	계란말이
	broth anchovy broth	멸칫국물	166	Kimchi Jjigae	김치찌개
			178	Kalguksu	칼국수
			194	Miyeokguk	미역국
			198	Tteokguk	떡국

	Ingredients		Page	Dishes	
B	broth anchovy broth	멸칫국물	200	Bulgogi	불고기
	broth beef broth	고깃국물	166	Kimchi Jjigae	김치찌개
			194	Miyeokguk	미역국
			198	Tteokguk	떡국
			200	Bulgogi	불고기
	broth vegetable broth	채소 국물	166	Kimchi Jjigae	김치찌개
			178	Kalguksu	칼국수
			194	Miyeokguk	미역국
			198	Tteokguk	떡국
			200	Bulgogi	불고기
	burdock root	우엉	170	Gimbap	김밥
	butter	버터	174	Toast (j-hope's)	제이홉표 토스트
			175	Toast (Jung Kook's)	정국표 토스트
C	cabbage	양배추	186	Dakgalbi	닭갈비
	carrots	당근	170	Gimbap	김밥
			178	Kalguksu	칼국수
			196	Japchae	잡채
			200	Bulgogi	불고기
			208	Gyeran Mari	계란말이
	cheese (sliced)	슬라이스 치즈	175	Toast (Jung Kook's)	정국표 토스트
			208	Gyeran Mari	계란말이
	cheongju (refined rice wine)	청주	186	Dakgalbi	닭갈비
			190	Jeyuk Bokkeum	제육볶음
			200	Bulgogi	불고기
			208	Gyeran Mari	계란말이
	chicken boneless chicken thighs	닭고기 다리 살	186	Dakgalbi	닭갈비
	chili peppers	고추	170	Gimbap (with Pork Belly)	삼겹살 김밥
	chives	부추	178	Kalguksu	칼국수
			182	Jjajang Ramyeon	짜장 라면
			196	Japchae	잡채
			200	Bulgogi	불고기

	Ingredients		Page	Dishes	
C	chives	부추	**206**	Kimchijeon	김치전
			208	Gyeran Mari	계란말이
	clams	조개	**178**	Kalguksu	칼국수
			194	Miyeokguk	미역국
			198	Tteokguk	떡국
	cucumber	오이	**170**	Gimbap	김밥
D	*doenjang* (soybean paste)	된장	**170**	Gimbap	김밥
E	eggs	계란	**164**	Ganjang Gyeran Bap	간장계란밥
			170	Gimbap	김밥
			174	Toast (j-hope's)	제이홉표 토스트
			175	Toast (Jung Kook's)	정국표 토스트
			198	Tteokguk	떡국
			200	Bulgogi	불고기
			208	Gyeran Mari	계란말이
F	fermented anchovy sauce	멸치 액젓	**178**	Kalguksu	칼국수
			194	Miyeokguk	미역국
			198	Tteokguk	떡국
			200	Bulgogi	불고기
	fish sauce	피쉬 소스	**178**	Kalguksu	칼국수
			194	Miyeokguk	미역국
			198	Tteokguk	떡국
			200	Bulgogi	불고기
	flour	밀가루	**178**	Kalguksu	칼국수
			206	Kimchijeon	김치전
G	garlic	마늘	**221**	Beef Broth	고깃국물
	garlic (minced)	다진 마늘	**166**	Kimchi Jjigae	김치찌개
			178	Kalguksu	칼국수
			186	Dakgalbi	닭갈비
			190	Jeyuk Bokkeum	제육볶음
			194	Miyeokguk	미역국
			196	Japchae	잡채
			198	Tteokguk	떡국
			200	Bulgogi	불고기
	gim (dried seaweed)	김	**170**	Gimbap	김밥
			198	Tteokguk	떡국
	ginger (minced)	다진 생강	**186**	Dakgalbi	닭갈비
	ginger powder	생강 가루	**190**	Jeyuk Bokkeum	제육볶음
	glass noodles (*dangmyeon*)	당면	**186**	Dakgalbi	닭갈비

	Ingredients		Page	Dishes	
G	glass noodles (*dangmyeon*)	당면	**196**	Japchae	잡채
			200	Bulgogi	불고기
	gochujang (Korean red chili paste)	고추장	**170**	Gimbap	김밥
			186	Dakgalbi	닭갈비
			190	Jeyuk Bokkeum	제육볶음
	gochutgaru (Korean chili powder)	고춧가루	**178**	Kalguksu	칼국수
			190	Jeyuk Bokkeum	제육볶음
H	ham	햄	**170**	Gimbap	김밥
			208	Gyeran Mari	계란말이
	honey	꿀	**186**	Dakgalbi	닭갈비
			190	Jeyuk Bokkeum	제육볶음
	hot chili pepper	청양고추	**178**	Kalguksu	칼국수
			186	Dakgalbi	닭갈비
I	instant *jjajang ramyeon* (instant noodles in black bean sauce)	인스턴트 짜장 라면	**182**	Jjajang Ramyeon	짜장 라면
J	jam fruit jam	과일잼	**174**	Toast (j-hope's)	제이홉표 토스트
	jam stawberry jam	딸기잼	**174**	Toast (j-hope's)	제이홉표 토스트
K	kelp (dried)	건조 다시마	**219**	Vegetable Broth	채소 국물
			220	Anchovy Broth	멸칫국물
	ketchup	케첩	**175**	Toast (Jung Kook's)	정국표 토스트
			208	Gyeran Mari	계란말이
	kimchi	김치	**166**	Kimchi Jjigae	김치찌개
			198	Tteokguk	떡국
			206	Kimchijeon	김치전
	kimchi juice	김칫국물	**166**	Kimchi Jjigae	김치찌개
			206	Kimchijeon	김치전
	kiwis	키위	**214**	Hwachae	화채
	Korean pancake mix	부침가루	**206**	Kimchijeon	김치전
	Korean radish	무	**219**	Vegetable Broth	채소 국물
			221	Beef Broth	고깃국물
	Korean zucchini	애호박	**178**	Kalguksu	칼국수
L	large green onion (*daepa*)	대파	**166**	Kimchi Jjigae	김치찌개
			178	Kalguksu	칼국수
			186	Dakgalbi	닭갈비
			190	Jeyuk Bokkeum	제육볶음
			198	Tteokguk	떡국
			200	Bulgogi	불고기
			219	Vegetable Broth	채소 국물

	Ingredients		Page	Dishes	
L	large green onion (*daepa*)	대파	**221**	Beef Broth	고깃국물
	lychees	리치	**214**	Hwachae	화채
M	*mandu* (dumpling)	만두	**198**	Tteokguk	떡국
	milk	우유	**174**	Toast (j-hope's)	제이홉표 토스트
			214	Hwachae	화채
	mozzarella cheese	모차렐라 치즈	**186**	Dakgalbi	닭갈비
			206	Kimchijeon	김치전
	mullyeot (corn syrup)	물엿	**170**	Gimbap	김밥
			186	Dakgalbi	닭갈비
	mushrooms	버섯	**178**	Kalguksu	칼국수
			208	Gyeran Mari	계란말이
			219	Vegetable Broth	채소 국물
	mushrooms button mushrooms	양송이버섯	**182**	Jjajang Ramyeon	짜장 라면
			196	Japchae	잡채
			198	Tteokguk	떡국
	mushrooms golden oak mushrooms	표고버섯	**196**	Japchae	잡채
			198	Tteokguk	떡국
			200	Bulgogi	불고기
	mushrooms golden oak mushrooms (dried)	건조 표고버섯	**219**	Vegetable Broth	채소 국물
	mushrooms oyster mushrooms	느타리버섯	**166**	Kimchi Jjigae	김치찌개
			196	Japchae	잡채
			200	Bulgogi	불고기
	mushrooms wood ear mushrooms	목이버섯	**196**	Japchae	잡채
	mussels	홍합	**178**	Kalguksu	칼국수
N	napa cabbage leaves	배춧잎	**200**	Bulgogi	불고기
O	oligo syrup	올리고당	**186**	Dakgalbi	닭갈비
	onion	양파	**166**	Kimchi Jjigae	김치찌개
			170	Gimbap	김밥
			178	Kalguksu	칼국수
			182	Jjajang Ramyeon	짜장 라면
			186	Dakgalbi	닭갈비
			190	Jeyuk Bokkeum	제육볶음
			196	Japchae	잡채
			200	Bulgogi	불고기
			206	Kimchijeon	김치전
			208	Gyeran Mari	계란말이
			219	Vegetable Broth	채소 국물
	oyster sauce	굴소스	**186**	Dakgalbi	닭갈비

	Ingredients		Page	Dishes	
O	oysters	굴	**194**	Miyeokguk	미역국
			198	Tteokguk	떡국
P	pasta noodles (linguine)	파스타 면(링귀니)	**178**	Kalguksu	칼국수
	pasta shells (conchiglie)	파스타 셸(콘킬리에)	**186**	Dakgalbi	닭갈비
	peach juice (canned)	복숭아 통조림 물	**214**	Hwachae	화채
	peaches (canned)	복숭아 통조림	**214**	Hwachae	화채
	peperoncino	페페론치노	**186**	Dakgalbi	닭갈비
	perilla leaves (*kkaennip*)	깻잎	**186**	Dakgalbi	닭갈비
			208	Gyeran Mari	계란말이
	perilla oil	들기름	**194**	Miyeokguk	미역국
	perilla powder	들깻가루	**194**	Miyeokguk	미역국
	pickled radish	단무지	**170**	Gimbap	김밥
	pineapple	파인애플	**214**	Hwachae	화채
	pollack (dried)	황태	**194**	Miyeokguk	미역국
	pork belly	**돼지고기** 삼겹살	**166**	Kimchi Jjigae	김치찌개
			170	Gimbap (with Pork Belly)	삼겹살 김밥
			190	Jeyuk Bokkeum	제육볶음
			206	Kimchijeon	김치전
	pork front hock	**돼지고기** 앞다리 살	**190**	Jeyuk Bokkeum	제육볶음
	pork shoulder	**돼지고기** 목살	**190**	Jeyuk Bokkeum	제육볶음
	potatoes	감자	**186**	Dakgalbi	닭갈비
R	*ramyeon* noodles (instant noodles)	라면 사리	**186**	Dakgalbi	닭갈비
	red chili pepper	홍고추	**186**	Dakgalbi	닭갈비
	rice	밥	**164**	Ganjang Gyeran Bap	간장계란밥
			170	Gimbap	김밥
			200	Bulgogi	불고기
	rice (uncooked)	쌀	**218**	Rice	밥
	rice cake (bite-sized)	떡볶이 떡	**186**	Dakgalbi	닭갈비
			200	Bulgogi	불고기
	rice cake (sliced)	떡국떡	**198**	Tteokguk	떡국
	rice flour for noodles	쌀가루(제면용)	**178**	Kalguksu	칼국수
	rice vermicelli	버미셀리 쌀국수	**196**	Japchae	잡채
S	scallions	쪽파	**164**	Ganjang Gyeran Bap	간장계란밥
			178	Kalguksu	칼국수
			182	Jjajang Ramyeon	짜장 라면
			196	Japchae	잡채
			200	Bulgogi	불고기
			206	Kimchijeon	김치전

	Ingredients		Page	Dishes	
S	scallions	쪽파	208	Gyeran Mari	계란말이
	scallops	가리비	178	Kalguksu	칼국수
	seaweed	미역	194	Miyeokguk	미역국
	shrimp	새우	196	Japchae	잡채
			206	Kimchijeon	김치전
	soda	사이다	214	Hwachae	화채
	soju	소주	186	Dakgalbi	닭갈비
			190	Jeyuk Bokkeum	제육볶음
			200	Bulgogi	불고기
	soy sauce	간장	164	Ganjang Gyeran Bap	간장계란밥
			170	Gimbap	김밥
			178	Kalguksu	칼국수
			186	Dakgalbi	닭갈비
			190	Jeyuk Bokkeum	제육볶음
			196	Japchae	잡채
			198	Tteokguk	떡국
			200	Bulgogi	불고기
	soy sauce for soup	국간장	194	Miyeokguk	미역국
			198	Tteokguk	떡국
			200	Bulgogi	불고기
	squid	오징어	196	Japchae	잡채
			206	Kimchijeon	김치전
	ssamjang (red chili and soybean paste)	쌈장	170	Gimbap (with Pork Belly)	삼겹살 김밥
	strawberries	딸기	214	Hwachae	화채
	sugar	설탕	166	Kimchi Jjigae	김치찌개
			170	Gimbap	김밥
			178	Kalguksu	칼국수
			186	Dakgalbi	닭갈비
			190	Jeyuk Bokkeum	제육볶음
			196	Japchae	잡채
			198	Tteokguk	떡국
			200	Bulgogi	불고기
			212	Dalgona	달고나
			214	Hwachae	화채
	sweet potatoes	고구마	186	Dakgalbi	닭갈비
T	toasted sesame oil	참기름	164	Ganjang Gyeran Bap	간장계란밥
			170	Gimbap	김밥
			178	Kalguksu	칼국수

	Ingredients		Page	Dishes	
T	toasted sesame oil	참기름	**186**	Dakgalbi	닭갈비
			190	Jeyuk Bokkeum	제육볶음
			194	Miyeokguk	미역국
			196	Japchae	잡채
			198	Tteokguk	떡국
			200	Bulgogi	불고기
	toasted sesame seeds	참깨	**164**	Ganjang Gyeran Bap	간장계란밥
			178	Kalguksu	칼국수
			190	Jeyuk Bokkeum	제육볶음
			196	Japchae	잡채
	tofu (bean curd)	두부	**164**	Ganjang Gyeran Bap	간장계란밥
			166	Kimchi Jjigae	김치찌개
V	vegan butter	비건 버터	**182**	Jjajang Ramyeon	짜장 라면
	vinegar	식초	**166**	Kimchi Jjigae	김치찌개
W	watermelon	수박	**214**	Hwachae	화채
	white wine (dry)	화이트 와인(드라이)	**186**	Dakgalbi	닭갈비
			190	Jeyuk Bokkeum	제육볶음
			200	Bulgogi	불고기
	wide glass noodles (wide *dangmyeon*)	넓적한 당면	**196**	Japchae	잡채
Z	zucchini	주키니	**178**	Kalguksu	칼국수

References

1 Easy Homemade Meals 간편한 집밥

Ganjang Gyeran Bap (Egg Rice with Soy Sauce) 간장계란밥

- Weverse | <BTS BON VOYAGE Season 4> EP. 8: Our Glowing Lights (서로가 본 서로의 빛)

Kimchi Jjigae (Kimchi Stew) 김치찌개

- YouTube | <Run BTS!> EP. 103: Avatar Chef King 2 (아바타 요리왕 2)
- Weverse | <BTS BON VOYAGE Season 4> EP. 4: Winter Finds Its Way in September (9월에 찾아온 겨울)

2 Box Lunches Just for You 취향 따라 만드는 도시락

Gimbap (Seaweed Rice Rolls) 김밥

- Weverse | Weverse Live (2020. 06. 21) - Today, we're gimbap chefs (오늘은 김밥 요리사)
- Weverse | Weverse Live (2020. 06. 21) - Today, we're really gimbap chefs (오늘은 내가 진짜 김밥 요리사)
- Weverse | <IN THE SOOP BTS ver. Season 1> EP. 6: Together, And On Their Own (다 같이 또 따로)

Toast (j-hope's & Jung Kook's Egg Sandwiches) 토스트

- Weverse | <IN THE SOOP BTS ver. Season 2> EP. 3:
  The Most Beautiful Moment in Life, Around the Table Together (화양연화, 함께하는 식탁)
- YouTube | [VLOG] j-hope (2019. 10. 03)
- Weverse | <IN THE SOOP BTS ver. Season 2> EP. 5: Permission to BTS
- YouTube | <Run BTS!> EP. 132: 77-Minute Debate 2 (77분 토론 2)

3 Slurp! Noodle Dishes 후루룩 면 요리

Kalguksu (Noodle Soup) 칼국수

- Weverse | <IN THE SOOP BTS ver. Season 2> EP. 3:
  The Most Beautiful Moment in Life, Around the Table Together (화양연화, 함께하는 식탁)
- YouTube | [BTS 진 with 백종원] Jin's Traditional Alcohol Journey (취중진담) EP. 4
- YouTube | [EPISODE] 지민 (Jimin) 'Set Me Free Pt.2' MV Shoot Sketch - BTS (방탄소년단)
- YouTube | <Run BTS!> EP. 142: The Perfect Combination (환상의 궁합)

Jjajang Ramyeon (Instant Noodles in Black Bean Sauce) 짜장 라면

- Weverse | <BTS BON VOYAGE Season 4> EP. 7: Happy Days with Seven of Us (일곱 명의 행복한 시간)
- YouTube | <Run BTS!> EP. 151: War of Money Hotel Staycation 2 (쩐의 전쟁 호캉스 2)
- Weverse | <IN THE SOOP BTS ver. Season 1> EP. 1: Excited to Begin (설레는 첫 출발)

4 Spicy Meat Dishes 매콤한 고기 요리

Dakgalbi (Spicy Stir-Fried Chicken) 닭갈비

- Weverse | <IN THE SOOP BTS ver. Season 2> EP. 5: Permission to BTS

Jeyuk Bokkeum (Spicy Stir-Fried Pork) 제육볶음

- YouTube | <Run BTS!> EP. 102: Avatar Chef King 1 (아바타 요리왕 1)

5 Dishes for Special Feasts 특별한 만찬

Miyeokguk (Seaweed Soup) 미역국

- Weverse | Weverse (2022. 10. 15) - 부산왔다아인교
- Weverse | Weverse (2021. 09. 13) - A Bit Late Happy Birthday (조금 늦은 생일축하)
- YouTube | [2020 FESTA] BTS (방탄소년단) '방탄생파' #2020BTSFESTA
- YouTube | [EPISODE] BTS (방탄소년단) @ "Yet To Come" in BUSAN

Japchae (Stir-Fried Glass Noodles and Vegetables) 잡채

- YouTube | [2020 FESTA] BTS (방탄소년단) '방탄생파' #2020BTSFESTA

Tteokguk (Sliced Rice Cake Soup) 떡국

- YouTube | <Run BTS!> EP. 44: Satisfy the Five Senses 2 (오감만족 2편)
- YouTube | BTS (방탄소년단) 7-Second Interview
- YouTube | BTS (방탄소년단) 2023 Happy Seollal Greeting

Bulgogi 불고기

- YouTube | [BTS VLOG] j-hope | 확신의 J VLOG
- YouTube | [BTS 진 with 백종원] Jin's Traditional Alcohol Journey (취중진담) EP. 4

6 Perfect Pairings with Drinks 술과 어울리는 안주

Kimchijeon (Kimchi Pancake) 김치전

- Weverse | <IN THE SOOP BTS ver. Season 2> EP. 4: BTS UNIVERSE
- YouTube | <Run BTS!> EP. 128: HELLO 2021
- YouTube | BTS (방탄소년단) '찐 방탄회식' #2022BTSFESTA

Gyeran Mari (Rolled Omelet) 계란말이

- YouTube | <Run BTS!> EP. 102: Avatar Chef King 1 (아바타 요리왕 1)

7 Sweet Desserts 달달한 디저트

Dalgona (Sugar Candy) 달고나

- YouTube | <Run BTS!> EP. 95: Let's Play, BTS 1 (방탄아 놀자 1)

Hwachae (Punch) 화채

- Weverse | <IN THE SOOP BTS ver. Season 2> EP. 2: Fly to "In the SOOP" with BTS (BTS가 IN THE SOOP을 여행하는 법)
- Weverse | Weverse (2020. 12. 28) - I can't believe they made me a snowman (내가 눈사람이라니)
- YouTube | <Run BTS!> EP. 131 77-Minute Debate 1 (77분 토론 1)
- YouTube | <Run BTS!> EP. 93 BTS Marble 1 (BTS 마블 1)

Book of Tasty Stories

BTS
RECIPE BOOK 2

GLOBAL EDITION

Publication Date	2023. 05. 10
Publisher	이충희 Lee, Choong Hee
IP Business Dept. Lead	최영남 Choi, Young Nam
IP Content Team Lead	서수진 Seo, Su Jin
Edu Content 1 Part Lead	정희은 Jeong, Hee Eun
Project Manager	정서경 Chung, Seo Kyung
Project Assistant	방수영 Bahng, Soo Yung
Marketing	이서진 Lee, Seo Jin (Lead)
	김근영 Kim, Kun Young
Global Biz Team	정석교 Jeong, Seok Kyo (Lead)
	박성희 Park, Sung Hee
Recipe Development, Food Styling, Photography	문스타 라이프 스튜디오 Moonstar Life Studio
English Translations	LEXCODE Inc.
English Proofreading	Patrick Anthony Ferraro
Design & Editing	스튜디오 닷츠 Studio Dots

Cake Corporation (Publication Registration May 24, 2022 No.2022-000170)
Pangyo TechONE Tower 1, 9F, 131, Bundangnaegok-ro,
Bundang-gu, Seongnam-si, Gyeonggi-do, 13529, Korea
ISBN 979-11-90996-61-7